Taguchi on Robust Technology Development

ASME PRESS SERIES ON INTERNATIONAL ADVANCES IN DESIGN PRODUCTIVITY

Taguchi on Robust Technology Development

Bringing Quality Engineering Upstream

by Genichi Taguchi
translated by Shih-Chung Tsai

ASME Press ■ New York ■ 1993

Originally published as *Quality Engineering for Technology Development* in 1990 by the Central Japan Quality Control Association.

Library of Congress Cataloging-in-Publication Data

Taguchi, Gen'ichi, 1924–
      Taguchi on robust technology development: bringing quality
engineering upstream/ Genichi Taguchi; translated by Tsai, Shih-Chung.
      p. cm.
      Includes bibliographic references and index.
      1. Taguchi methods (Quality control)   I. Title

      TS156.T342   1993
      658.5'62 – dc20   92-30947
                                            CIP

ISBN   0-7918-0028-8

Contents

Editor's Note

This book defines and amplifies the emerging science now called *quality engineering*. Several companies (including AT&T, Xerox, Ford and ITT) have applied Dr. Taguchi's work in the various phases of product realization. Thousands of case studies are now available in industry that demonstrate the power of the Taguchi system of quality engineering. It is now clearly understood that quality cannot be inspected into a product or a process. In this important work, Dr. Taguchi demonstrates the applicability of his methods to the earliest phases of the design process. The reader will learn how to develop robust technologies. Since, generally, 80 percent of the cost of a product is determined by decisions made in the first 20 percent of the design process, it is important to build quality into technologies that will at some point be employed in new products. This implies a sequence of activities including identification of potential noise factors, control and tuning factors, and appropriate performance criteria, which can be used (through experimentation or simulation) to maximize the "functionability" of a new or traditional technology. This will allow development of high-quality, low-cost products in a timely fashion. In fact, the techniques presented offer the opportunity to make "time to market" an input rather than an outcome.

Kenneth M. Ragsdell

Translator's Note

D<small>r.</small> Genichi Taguchi's off-line quality control is a very efficient tool for developing high-quality products at a low cost. The theme of off-line quality control is to design robust products that can withstand both downstream production and disturbances due to usage. It does not, however, attempt to control the sources of downstream disturbances; hence, it is much more cost efficient than traditional on-line quality control. In this book, Dr. Taguchi brings off-line quality control concepts upstream to the technology development stage.

The keynote of this new approach is to design robustness into a generic technology so that the new technology can perform its intended functions under realistic downstream production or usage conditions. After one generic technology is developed and matures, the accumulated technological know-how can be applied repetitively and flexibly to develop a range of new products. Thus, in today's quick-changing market, this approach will be more time and cost efficient in meeting various customers' needs. Consequently, the newly developed technology can increase both the productivity and profitability of a company by reducing development cost and time to market. In the next century, the capability of developing robust technology will be essential to the competitiveness of any manufacturing enterprise.

Shih-Chung Tsai

Dr. Shih-Chung Tsai specializes in quality engineering and Taguchi-type experimental design and is an instructor and consultant of experimental process using robust design in the Department of Engineering Processes at General Motors Corporation, Rochester Division.

Foreword

Since its introduction into the United States in 1980, Dr. Taguchi's system of quality engineering has been widely applied and broadly discussed. Implementation has suffered from the blind-men-and-the-elephant syndrome. Most people who have encountered this comprehensive system of quality engineering have grabbed on to some part of it and said, "Now I understand what this elephant is." Also, there have been contextual problems. Many researchers have complained that it does not seem similar to research, which is not surprising since it is intended for development and production, not research.

In this book Dr. Taguchi has set forth in simple, concise terms the full, comprehensive system and has clearly shown its role in developing and producing new products. Even more importantly, he has described the role of quality engineering in providing flexible technologies that will enable the rapid development of the product variety that is the new competitive playing field.

Dr. Taguchi's comprehensive system of quality engineering is one of the great engineering achievements of the twentieth century. This book will help communicate the full scope and context of this system to workers in product development and production. This book is required reading for the product people in any corporation that hopes to remain competitive.

Don Clausing

Preface

The term *Taguchi methods* was coined in the United States. It pertains to the evaluation and improvement of the robustness of products, tolerance specifications, the design of engineering management processes, and the evaluation of the economic loss caused by the functional variation of products. However, I prefer the term *quality engineering* instead. The purpose of this book is to explain just what quality engineering is.

In this high-tech generation, it is very important for any enterprise to develop robust technologies that can efficiently develop new products, refine current ones, and manufacture products flawlessly in terms of quality and cost. This book is written for administrators, managers, technology developers, and engineers of manufacturing enterprises to help meet this challenge. It also contains a general discussion about the productivity of manufacturing enterprises, which is equivalent to cost reduction and quality improvement at the following five manufacturing stages:

1. technology development
2. product planning
3. product design
4. design of the production process
5. management of the production process

In addition to an economic evaluation of the quality of products, I present several easy-to-understand examples to show managers and engineers how to develop technologies, design products, and invest in process design and process management more economically and meaningfully than before. In the United States, economic evaluation of products is usually considered to be a management affair; however, I consider it a technical affair. The methods of applying the economic evaluation of products to engineering problems and improving the quality of products at low cost by parameter design are the major themes of this book.

This book can be used for a twelve-hour course on quality engineering. The exercises in each chapter illustrate how to solve different types of quality engineering problems that actually arise. For a one-day course (six or seven hours), engineers or technicians from manufacturing departments can refer to Chapters 1, 2, and 4; those from design departments may refer to Chapters 1, 3, and 5.

For a more basic treatment of the subject, I suggest the seven-volume *Quality Engineering*, particularly volumes 1 and 2 (Japanese editions are available from the Japanese Standard Association, and English editions will be published soon.). Finally, I hope that administrators, managers, and engineers of manufacturing enterprises understand the importance of technology development, product design, and process management after reading this book.

Genichi Taguchi

1

Quality and Productivity

This chapter illustrates what is meant by *quality* and describes the duties of manufacturers to improve quality, emphasizing the roles played by design departments.

1.1 PRODUCT PLANNING AND QUALITY

Generally enterprises first plan what new products to introduce. Second, they design the processes to manufacture these products. Finally, they actually produce and sell the products. Although the problems of quality in services are similar to those in material goods, this book will focus on goods.

The major stages of product development are as follows:

1. *Product planning.* Market analysis is conducted, the objective functions (primary goals) of the product are determined, pricing is set, and the lifespan of the new product is decided.
2. *Product design.* The product is designed and developed.
3. *Production process design.* The process for manufacturing the product is designed and developed.
4. *Production.* The product is manufactured.
5. *Sale.* The product is sold.
6. *Product service.* After-sale service matters are handled, such as warranties and claims of defective products.

In general, the selling price of a product is usually several times its manufacturing cost. To develop new products, one must first predict the prices of competitive products that may appear in the marketplace. Then, the design department is given a deadline by which it must design a new product that can be produced at a cost that is a fraction of the proposed sale price. Therefore, the competitiveness of a manufacturing enterprise depends on its ability to design and develop new products that can perform the desired functions, which are specified at the product planning stage, while keeping production costs below those for competitive products.

In this book, quality is evaluated by quality loss, defined as the amount of functional variation of products plus all possible negative effects, such as environmental damages and operational costs. The objective of this book is to illustrate the following methods for improving the quality of products:

> ► evaluation methods for determining the quality loss of products (functional variation of products plus the cost caused by the damaging effects of defective products)
> ► parameter design methods for improving quality level (or reducing quality loss) without increasing the cost of products
> ► tolerance design methods for balancing the tradeoff between the cost of high-grade components and the total quality loss of products
> ► optimal quality management methods for controlling the objective characteristics of products during the manufacturing process

These methods compose quality engineering, or, as it is known in the United States, "quality engineering for low cost." Several decades ago, low-quality Japanese products were not competitive, although they were very cheap. However, in the current global market, high-quality Japanese products are losing their competitiveness because they are becoming increasingly expensive. Only inexpensive but high-quality products can survive in the highly competitive global market of today and tomorrow.

At the first stage of product planning, many manufacturing enterprises specify target values for the manufacturing costs and reliability of new products. However, it is not a good idea that these target values be decided at this stage; rather, only the target values of the objective functions, the selling price, and the operational life of a new product should be determined at this first stage.

Consequently, setting the target values for the cost and quality level of new products should be the responsibility of the technology and production departments, not the product planning department. Unfortunately, few enterprises allot enough of their budgets to encourage their design and technology departments to conduct robust (that is, reliable) technology development to ensure that new products can perform the planned functions or targets and still be manufactured at low cost.

Simply put, the purpose of quality engineering is to conduct the research necessary to develop robust technologies and methods that increase the competitiveness of new products by reducing their cost and improving their quality; this enables the manufacturing enterprise to survive in the highly competitive global market. In addition, it is very important that the staffs of product design, production process design, and quality management departments reduce the cost and improve the quality of new products as much as possible before a specified deadline at the stage of new product design and development. One can never know in advance what the quality levels and production costs of competing products will be. Thus, one is at a competitive disadvantage no matter how well the developed products match the planned targets if their cost and quality levels do not compare favorably with those of the competition. Conclusively, the targets should not be decided for any kind of quality characteristic during new product planning. Instead, it is the duty of design departments to specify the targets for quality characteristics.

Ideally, product design departments should design new products that have no manufacturing cost, no malfunctions, and no damaging environmental effects. To approach these ideal goals, the targets of quality and cost that design departments are asked to achieve should not be set up at the stage of new product planning; instead, these targets should be developed by design departments themselves. In this way design departments will not be restricted to achieving specific targets, and they can try their best to improve the quality level and reduce the cost of the new products as much as possible.

1.2 THE DUTIES OF DESIGN ENGINEERS AND PRODUCTION TECHNICIANS

Both product design and production process design can be divided into the following five stages:

1. *System selection.* First, one must consider all possible systems that can perform the required functions. In addition to existing systems, new systems should be considered that have not yet been developed. The advantage of developing new systems is that they are usually protected by patents. In the United States, everyone who might contribute to the development of new systems usually gets together in brainstorming sessions to discuss all possible strengths and weaknesses of new proposed systems. New systems are then chosen based on the judg-

ments and discussions of these people. Although all such judgments and discussions are based on qualitative rather than quantitative criteria, they are very important in selecting new systems. If design departments have enough time, financing, and other resources, they should develop several new systems simultaneously.

2. *Parameter design.* At this stage, design engineers should specify appropriate design parameters (system parameters) of the chosen systems to improve quality and reduce cost. The nonlinear relations among system parameters and the interactions among system parameters and "noise," or, environmental, factors have been widely applied to desensitize the objective functions of new systems against various factors. Right now, in addition to experimental design methods, response surface linear programming, nonlinear programming, and operating window methods have made parameter design more efficient.

3. *Tolerance design.* After parameter design, one must decide the tolerance specifications for all components of the chosen systems and choose appropriate grades of materials for these components. The objective of tolerance design is to decide the tradeoff between quality level and cost in designing new systems. The objective evaluation criteria of tolerance design are the quality losses of products. These quality losses are usually estimated by the functional deviation of the products from their target values plus the cost due to the malfunction of these products.

4. *Tolerance specifications.* After tolerance design, the grades of the materials and the tolerance limits of the components must be specified and placed on blueprints, and contracts with material and component suppliers must

be signed. It is very common to combine the tolerance specification stage with the tolerance design stage.

5. *Quality management for the production process.* At this stage, feedback control systems must be designed to control the statistical distribution of the critical dimensions of new products. The design of gauges for measuring quality for use by the quality management system is especially important at this stage. Production design engineers and technicians should determine the best quality management methods that balance the quality level of the products and the cost of high-grade components and materials.

Of the five stages, system selection is most relevant to the designers' creativity and the technologies that have been developed. In actual applications, it usually takes a few days to select appropriate systems for new products. Generally, new systems might not be selected until stages 2 or 3. However, quality engineering is more related to stages 2 through 5 than to stage 1.

Compared with product design or process design, the development of new products can be broken down into the following steps:

A. system design
B. subsystem design
C. element or component design
D. development of elements or components
E. development of raw materials

Generally, each of these steps can be conducted through the five product (or process) design stages described above. It is very common that in steps C through E, product design and production process design are conducted simultaneously.

1.3 THE DIFFERENCES BETWEEN SCIENCE AND ENGINEERING _____

The differences between science and engineering are very great. The goal of scientific research is to describe the principles governing natural phenomena as clearly as possible. In other words, the objective of scientific research is to find the best way to understand natural phenomena without regard for cost. Therefore, this approach can be very expensive.

In the world of engineering, however, there might be many ways to design and manufacture products so that they can perform particular functions. From all possible approaches, the best one is chosen by considering the quality level and the cost required to develop very competitive products; otherwise, one will not be able to survive in the highly competitive market. In the world of science, usually only one mathematical formula or principle best describes the objective characteristic of the phenomenon being studied. In general, financial cost will not be a major issue in scientific research.

Rather than scientific research, this book will focus only on engineering and will illustrate how to apply experimental design methods to design functional robustness into products and technologies. A design is said to be "functionally robust" if it inherently tends to diminish the effect of input variation on performance. Therefore, the purposes of the experimental design methods applied in quality engineering are quite different from those of scientific experimental design methods. Most statistical textbooks and experimental design textbooks are written by statisticians for scientific research. Thus, these textbooks apply statistical experimental design to basic scientific research. In contrast, quality engineering books (including those written by this author) concerning technology development and product design are written especially for product design engineers, technology development engineers, and others involved in actual engineering design.

Deciding the function(s) of a new product is the duty of the product planning department. In fact, this task is not related to either science or engineering, but pertains to human, or social, factors. After deciding upon the objective function(s), one must develop a product that can perform the function(s) soundly and robustly under various operating conditions. Ensuring the functional robustness of these new products or technologies while keeping costs down is the major purpose of quality engineering. In summary, it is the objective of quality engineering to choose from all possible designs the one that can ensure the highest functional robustness of products at the lowest possible cost.

Because system selection is a very important task in quality engineering, one needs to consider all possible technologies available at this initial stage. After selecting appropriate new systems, one must then decide the nominal values for the parameters of new systems and then specify their tolerances. In quality engineering, the approaches to various types of problems can be quite different, as illustrated by the two approaches below.

Take the power supply network for television sets as an example. First, let the parameters of this network be A, B, and so on. After selecting the new system for this power supply network, one must decide the nominal values and tolerances for the components of this network. Assume that the objective characteristics of this network are the output voltage y and the output current z. The nominal values of y and z are decided by the parameters of the network system; thus, y and z are functions of these system parameters:

$$y = f(A, B, \ldots)$$
$$z = g(A, B, \ldots)$$

Let the target values of y and z be y_0 and z_0. If one can solve the following simultaneous equations, it will be very easy to determine the optimal nominal values for the parameters:

$$f(A, B, \ldots) = y_0$$
$$g(A, B, \ldots) = z_0$$

However, there might be an infinite set of solutions to these two simultaneous equations. In addition, the nominal values chosen for these two parameters will affect the quality level and cost of the power supply network. In quality engineering, the optimal solution to this problem is to set these system parameters so that the sum of quality loss and the cost of the network system are minimized. The optimal solution is also called the "optimal condition" or "optimal design" of the system parameters.

As a second example, consider the exhaust catalyst of an automobile engine. The objective characteristic of an exhaust catalyst y will be the conversion ratio of the damaging contents of the exhausted gas. It is common sense that the ideal conversion ratio is 100%. In other words, the target for an exhaust catalyst is to remove all the damaging contents of the exhausted gas. However, there may be no way to achieve this ideal condition. It is obvious that the quality engineering approach for the power supply example, which has an infinite set of solutions, and the approach for the catalyst, which has no ideal solution, are quite different.

Of course, there might be many sets of parameters that can make the conversion ratio very close to the target of 100%. Among all these sets, the engineer should choose the one that costs the least. In the example of the power supply network, one might need to apply different evaluation criteria (such as the quality loss function) to determine the optimal setting for the system parameters.

Clearly, at the stage of product development, one must develop a product that can perform the specified functions soundly and flawlessly. However, there may be many different ways to meet this requirement. Thus, one must consider the cost of these solutions as another evaluation criteria. In addition, one should not be influenced by misleading prejudices; examples include the belief that Japanese cars are all correctly designed and American cars are all incorrectly designed. In fact, both Japanese cars and American cars are correctly designed. To decide which ones are better designed, one should evaluate which ones are manufactured at lower cost, have fewer problems, make less noise, and achieve higher mileage. In other words, one should focus on quality and cost.

Let us call the sum of quality loss and production cost the *productivity loss*, and the inverse of *productivity loss* becomes the *productivity index*. The objective of research in technology development is to increase the productivity index of products as much as possible. The first five stages of product (or process) design described in Section 1.1 are very important factors in increasing the productivity index of new products. Since improving the quality level and reducing the cost of products are directly related to their production processes, production technology departments and production departments should focus their efforts on achieving these ends and thus improving the productivity index of these products.

However, modern technology in quality engineering is not intended to reduce the sources of variation in products directly. Instead, one needs to make the systems of products or production processes less sensitive to sources of uncontrollable noise, or outside influences, through parameter design (off-line quality control) methods. The reason is that the latter is much cheaper than the former. After parameter design, it is common to use tuning methods (on-line process control methods) to ensure that the objective characteristics of products meet target specifications. Chapter 4 deals with on-line quality control methods and Chapter 5 with parameter design (off-line quality control) methods.

1.4 SOURCES OF NOISE AND CORRESPONDING MANAGEMENT STRATEGIES

Variations in the objective functions of products (or technologies) are primarily due to three sources: environmental effects, deteriorative effects, and manufacturing imperfections. The purpose of robust design is to make the products and the processes less sensitive to these effects. The relationship between the three major sources of noise and the corresponding strategies of enterprises to deal with the noise is shown in Table 1.1.

The table indicates, for instance, that it should be the responsibility of process management (row 1 of manufacturing) to measure periodically the quality of products during the production process. The causes of defective products should be rooted out, and the production processes should be readjusted to normal conditions. In process management both feedforward (open-loop) control and feedback (closed-loop) control are widely used to achieve these purposes. It is then up to product management (row 2 of manufacturing) to periodically check the objective characteristics of products and the equipment used to measure these characteristics. For more details about these two strategies (process management and product management), refer to Chapter 4.

Since on-line quality control methods are widely used in the production processes of most manufacturing enterprises, production technology departments should be responsible for designing

TABLE 1.1   Sources of noise and the corresponding strategies for managing the noise in each department

Department	Strategy	Outer (environmental effects)	Inner (deteriorative effects)	Manufacturing Imperfections
Technology departments				
Development and design	Modify:			
	1. System design	R	R	R
	2. Parameter design	R	R	R
	3. Tolerance design	N	R	R
Production technology	Modify:			
	1. System design	X	X	R
	2. Parameter design	X	X	R
	3. Tolerance design	X	X	R
On-line departments				
Manufacturing	Modify:			
	1. Process management	X	X	R
	2. Product management	X	X	R
Marketing	Modify:			
	1. After-sale service	X	X	X

Noises spans the three rightmost columns (Outer, Inner, Manufacturing).

R: The effects of this type of noise can be reduced in this department using the strategy indicated.

N: Although the effects of this type of noise can be reduced, it is not recommended to do so at this stage.

X: The effects of this type of noise cannot be reduced in this department using the strategy indicated.

quality management systems. In many manufacturing enterprises, production technology departments work together with production departments to design these on-line quality control systems.

Exercise

[1.1]   Illustrate the ideal function of a spring. In addition, illustrate the differences between the objective functions of an ideal spring and those of a real spring. Consider the operational life of a spring in your answer.

2

Methods for Evaluating Quality

2.1 COMPARING THE QUALITY LEVELS OF SONY TV SETS MADE IN JAPAN AND IN SAN DIEGO

The front page of the *Ashi News* on April 17, 1979 compared the quality levels of Sony color TV sets made in Japanese plants and those made in the San Diego, California, plant. The quality characteristic used to compare these sets was the color density distribution, which affects color balance. Although all the color TV sets had the same design, most American customers thought that the color TV sets made in the San Diego plant were of lower quality than those made in Japan.

The distribution of the quality characteristic of these color TV sets was given in the *Ashi News* and is shown in Figure 2.1. As shown in Figure 2.1, the quality characteristics of the TV sets from Japanese Sony plants are normally distributed around the target value *m*. If a value of 10 is assigned to the range of the tolerance specifications for this objective characteristic, then the standard deviation of this normally distributed curve can be calculated and is about $^{10}/_6$.

In quality control, the process capability index (C_p) is usually defined as the tolerance specification divided by 6 times the standard deviation of the objective characteristic:

$$C_p = \frac{\text{Tolerance}}{6 * \text{Standard deviation}}$$

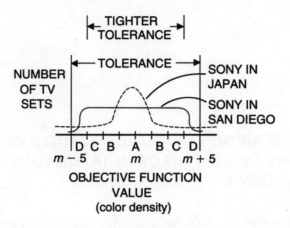

Figure 2.1 Objective function value distribution of color density in Sony TVs where *m* is the target value of the color density distribution; 10 is the range of tolerance specifications; and A, B, C and D are the "grades."

Therefore, the process capability index of the objective characteristic of Japanese Sony TV sets is about 1. In addition, the mean value of the distribution of these objective characteristics is very close to the target value *m*.

On the other hand, a higher percentage of the TV sets from San Diego Sony are within the tolerance limits than those from Japanese Sony. However, the color density of San Diego TV sets is uniformly distributed rather than normally distributed. Therefore, the standard deviation of these uniformly distributed objective characteristics is about $1/\sqrt{12}$ of the tolerance specification. Consequently, the process capability index of the San Diego Sony plant is calculated as follows:

$$C_p = \frac{\text{Tolerance}}{6\left(\dfrac{\text{Tolerance}}{\sqrt{12}}\right)} = 0.577$$

It is obvious that the process capability index of San Diego Sony is much lower than that of Japanese Sony.

All products that are outside of the tolerance specifications are supposed to be considered defective and not shipped out of the plant. Thus products that are within tolerance specifications are assumed to pass and are shipped. As a matter of fact, tolerance specifications are very similar to tests in school, where 60% is usually the dividing line between passing and failing, and 100% is the ideal score. In our example of color TV sets, the ideal condition is that the objective characteristic, color density, meet the target value m. The more the color density deviates from the target value, the lower the quality level of the TV set. If the deviation of the color density is over the tolerance specifications, $m \pm 5$, a TV set is considered defective. In the case of a school test, 59% or below is failing, while 60% or above is passing. Similarly, the grades between 60% and 100% in evaluating quality can be classified as follows:

60%–69%	Passing (D)
70%–79%	Fair (C)
80%–89%	Good (B)
90%–100%	Excellent (A)

The "grades" D, C, B, and A in parentheses above are quite commonly used in the United States for categorizing the objective characteristics of products. Thus, one can apply this scheme to the classification of the objective characteristic (color density) of these color TV sets as shown in Figure 2.1. (Please note that this classification scheme was not used in the report in the Ashi newspaper.) One can see that a very high percentage of Japanese Sony TV sets are within grade B, and a very low percentage are within or below grade D. In comparison, the color TV sets from San Diego Sony have about the same percentages in grades A, B, and C.

To reduce the difference in the process capability indices between Japanese Sony and San Diego Sony, (and thus seemingly increase the quality level of the San Diego sets) the latter tried to tighten the tolerance specifications to extend only to grade C shown in Figure 2.1 rather than grade D. Therefore, only the products within grades A, B, and C were treated as passing. But this approach is faulty. Tightening the tolerance specifications because of a low process capability in a production plant is as meaningless as increasing the passing score of school tests from 60% to 70% just because students do not learn well. On the contrary, such a school should consider asking the teachers to lower the passing score for students who do not test as well instead of raising it. The next section will illustrate how to evaluate the functional quality of products meaningfully and correctly.

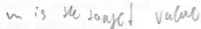

2.2 LOSS FUNCTION _____

When an objective characteristic y deviates from its target value m, some financial loss will occur. Therefore, the financial loss, sometimes referred to simply as the quality loss or used as an expression of quality level, can be assumed to be a function of y, which we shall designate $L(y)$. When y meets the target m, the loss of $L(y)$ will be at a minimum; generally, the financial loss can be assumed to be zero under this ideal condition:

$$L(m) = 0 \qquad (2.1)$$

Since the financial loss is at a minimum at this point, the first derivative of the loss function with respect to y at this point should also be zero. Therefore, one obtains the following equation:

$$L'(m) = 0 \qquad (2.2)$$

If one expands the loss function $L(y)$ through a Taylor series expansion around the target value m and takes Equations (2.1) and (2.2) into consideration, one will get the following equation:

$$L(y) = L(m) + \frac{L'(m)}{1!}(y - m) + \frac{L''(m)}{2!}(y - m)^2 + \cdots$$

$$= \frac{L''(m)}{2!}(y - m)^2 + \cdots$$

This result is obtained because the constant term $L(m)$ and the first-derivative term $L'(m)$ are both zero. In addition, the third-order and higher-order terms are assumed to be negligible. Thus, one can express the loss function as a squared term multiplied by a constant k:

$$L(y) = k(y - m)^2 \tag{2.3}$$

When the deviation of the objective characteristic from the target value increases, the corresponding quality loss will increase. When the magnitude of the deviation is outside of the tolerance specifications, this product should be considered defective. Let the cost due to a defective product be A and the corresponding magnitude of the deviation from the target value be Δ. Taking Δ into the right-hand side of Equation (2.3), one can determine the value for the constant k by the following equation:

$$k = \frac{\text{Cost of a defective product}}{(\text{Tolerance})^2} = \frac{A}{\Delta^2}$$

In the case of the Sony color TV sets, let the adjustment cost be $A = \$6$ when the color density is out of the tolerance specifications.

Therefore, the value of k can be calculated by the following equation:

$$k = \frac{600}{5^2} = \$0.24$$

Therefore, the loss function is

$$L(y) = \$0.24(y - m)^2$$

This equation is still valid even when only one unit of product is made. Consider the visitor to the BHEL Heavy Electrical Equipment Company in India who was told, "In our company, only one unit of product needs to be made for our nuclear power plant. In fact, it is not really necessary for us to make another unit of product. Since the sample size is only one, the variance is zero. Consequently, the quality loss is zero and it is not really necessary for us to apply statistical approaches to reduce the variance of our product."

However, the quality loss function $[L = k(y - m)^2]$ is defined as the mean square deviation of objective characteristics from their target value, not the variance of products. Therefore, even when only one product is made, the corresponding quality loss can still be calculated by Equation (2.3). Generally, the mean square deviation of objective characteristics from their target values can be applied to estimate the mean value of quality loss in Equation (2.3). One can calculate the mean square deviation from the target σ^2 (σ^2 in this equation is not the variance) by the following equation (the term σ^2 is also called the *mean square error* or the *variation of products*):

$$\sigma^2 = \text{mean value of } (y - m)^2$$

TABLE 2.1   Quality levels of Sony TV sets where the tolerance specification is 10 and the objective function data corresponds to Figure 2.1

Plant Location	Mean Value of Objective Function	Standard Deviation	Variation	Loss L (in $)	Defective[1] Ratio
Japan	m	$\dfrac{10}{6}$	$\dfrac{10^2}{36}$	0.667	0.27%
San Diego	m	$\dfrac{10}{\sqrt{12}}$	$\dfrac{10^2}{12}$	2.00	0.00%

[1]Defective ratio is percentage of products found with defects (here it means those scrapped).

Substituting this equation into Equation (2.3), one gets the following equation:

$$L = k\sigma^2 \qquad (2.4)$$

From Equation (2.4), one can evaluate the differences of average quality levels between the TV sets from Japanese Sony and those from San Diego Sony as shown in Table 2.1. From Table 2.1 it is clear that although the defective ratio of the Japanese Sony is higher than that of the San Diego Sony, the quality level of the former is 3 times higher than the latter. Assume that one can tighten the tolerance specifications of the TV sets of San Diego Sony to be $m \pm 10/3$. Also assume that these TV sets remain uniformly distributed after the tolerance specifications are tightened. The average quality level of San Diego Sony TV sets would be improved to the following quality level:

$$L = \$0.24\left[\left(\frac{1}{\sqrt{12}}\right)(10)\left(\frac{2}{3}\right)\right]^2 = \$0.889$$

where the value of the loss function is considered the relative quality level of the product. This average quality level of the TV sets of San Diego Sony is $1.11 higher than the original quality level but still $0.22 lower than that of Japanese Sony TV sets. In addition, in this type of quality improvement, one must adjust the products that are between the two tolerance limits, $m \pm {}^{10}/_3$ and $m \pm 5$, to be within $m \pm {}^{10}/_3$. In the uniform distribution shown in Figure 2.1, 33.3% would need adjustment, which would cost $6 per unit. Therefore each TV set from San Diego Sony would cost an additional $2.00 on average for the adjustment:

$$(6)(0.333) = \$2$$

Consequently, it is not really a good idea to spend $2 more to adjust each product, which is worth only $1.11. A better way is to apply quality management methods to improve the quality level of products, as described in Chapter 4.

In September 1980 a visitor at the River Plant of the GE Corporation noticed that the operators of the plant widely applied go/no-go type gauges to check whether their products were passing inspection or not. However, these operators did not realize the importance of tightening the variation of the objective characteristics of products to reduce quality loss. The visitor suggested that they should use Shewhart control charts to increase the quality level of their products. However, one agent in their quality assurance department said that the goal of their quality management was that their products have a zero defective ratio; in addition, the department would be satisfied if all the objective characteristics of their products were within tolerance specifications.

This quality management concept is very misleading. Very few Japanese companies would be satisfied with quality levels that were just within the tolerance specifications of the Japanese Industrial Standard. The reason is that most Japanese companies try to reduce the variations of the objective characteristics of their products even when these characteristics are already within tolerance specifications.

For example, many years ago, Nippondenso Co., Ltd., instructed the engineers in their production plants and supplier plants to increase their process capability index to be above 1.33. To meet this goal, the engineers took several samples of their production output every day and measured the deviation of the objective characteristics of these samples from the target values. They continued sampling and measuring the objective characteristics for three months. Assume that the measured data of the objective characteristics of these samples are $y_1, y_2, \ldots, y_n$. Then the standard deviation of the variation of these objective characteristics can be calculated by the following equation:

$$\sigma = \sqrt{\frac{1}{n}\left[(y_1 - m)^2 + (y_2 - m)^2 + \cdots + (y_n - m)^2\right]}$$

Therefore, the process capability index is

$$C_p = \frac{\text{Tolerance}}{6\sigma} \qquad (2.5)$$

From Equation (2.5), one also gets the loss function

$$L = k\sigma^2 \qquad (2.6)$$

Using Equations (2.5) and (2.6) as the evaluation criteria for production processes, engineers can increase the process capability index of their products more efficiently than by using a zero defective ratio.

27

Discussion: About Quality Loss Function

Q: Quality loss function is defined as the square of the deviation of an objective characteristic from its target. However, it is still difficult for me to understand the actual meaning of quality loss function. Please explain the differences between second-order quality loss function and the first-order quality loss of accumulation methods.

A: The reason for using the square deviation to approximate quality loss is based on consideration of all possible operational and noise conditions, which are usually randomly and normally distributed. In comparison, accumulation methods take into account the deviation from nominal specifications; however, the quality loss of accumulation methods is linearly proportional to the deviation of quality characteristics from their target. The financial loss calculated using accumulation methods is defined as below:

$$L_{\text{accumulation}} = \begin{cases} a(m - y) & \text{when } y < m \\ b(y - m) & \text{when } y > m \end{cases}$$

In this equation, m is the nominal value of an objective characteristic (e.g., the nominal weight of calcium per pound of infant formula), while the coefficient a is the unit financial loss for customers when the content of this product is less than m; the coefficient b is the unit financial loss for manufacturers when this characteristic exceeds its nominal value. Thus, the average financial loss L can be evaluated from sampling data $y_1, y_2, \ldots, y_n$ by the following equation:

$$L_{\text{accumulation}} = \frac{1}{n} \left[a\Sigma'(m - y) + b\Sigma''(y - m) \right] \qquad (2.7)$$

In this equation, Σ' is the summation of the y's that are smaller than m, and Σ'' is the summation of y's that are larger than m. In this example, we assume that increasing or decreasing this characteristic (e.g., the weight of calcium per pound of infant formula) will not affect the quality loss of the other characteristics (e.g., the content of vitamin A in the same infant formula). Otherwise, the quality loss in Equation (2.7) must be further adjusted.

Q: Does the quality loss function really approximate the effects of all possible operational conditions and noise conditions?

A: Yes, that is exactly the definition of quality loss function. This function approximates the actual quality loss by the second-order term in the Taylor series expansion. The constant k in this equation is usually decided by the financial loss A_0 and the corresponding LD50 point (the point at which the probability of failure (or falling outside tolerance specifications) is 50%).

Q: In actual application, the second-order term of the Taylor expansion seems to be appropriate in most cases. In comparison, the first-order term is not very precise, and the third-order term is not very practical in general applications. Therefore, isn't the second-order term the best approximation of the actual quality loss in most cases?

A: Generally, yes. In fact, this problem really depends on the objective characteristics of products. Basically, this equation was developed primarily from a consideration of the second-order expansion. This quality loss function is good for most applications. However, the answer to this question is controversial.

Exercise

[2.1]   A manager is going to buy one of the following three types of materials, A_1, A_2, or A_3, for a new product. The thermal expansion coefficient b (the expansion per degree Celsius) and the wear ratio per year β of these three materials are shown below:

	$b(\%)$	$\beta(\%)$	Price ($)
A_1	0.08	0.15	1.80
A_2	0.03	0.06	3.50
A_3	0.01	0.05	6.30

However, when the objective dimensions of the three types of materials deviate from their target by an amount of $\Delta_0 = 10\%$, malfunction will occur and the cost will be $A_0 = \$280$.

a. Assume that the standard deviation of the variation of the working temperature is $\sigma_{Temp} = 15°C$, and the expected operational life of this new product is 10 years. Of the three materials A_1, A_2, and A_3, which one is the best (optimal) choice?

b. Find the optimal tolerance specifications (in percent) for the initial dimension of the optimal material determined in (a).

    Hint: The total cost is calculated by summing the quality loss caused by the variations of the objective dimensions (manufacturing imperfections are not considered here) and the corresponding price, as shown in the following equation (in this equation, T is the expected operational life of the product):

$$L = \text{Price} + \frac{A_0}{\Delta_0^2}\left[\left(b^2\right)\left(\sigma_{Temp}^2\right) + \frac{T^2}{3}\beta^2\right]$$

3

Methods for Specifying Tolerances

Quality engineering is very much related to the quality loss function described in Chapter 2. However, in actual negotiations between vendors or between design and manufacturing engineers or in the quality management of components or products, the tolerances of components or products are specified more often than the quality loss function. This chapter explains how to specify tolerances for components and products at the design stage.

3.1 OBJECTIVE

In product management or trading negotiations, tolerance specifications are commonly employed to control the variances of objective characteristics of components or products. Safety factors are an important consideration in determining tolerance specifications. Recently simulation methods based on probability theory have been widely used to evaluate the failure ratio of products to estimate these safety factors. However, simulation methods are inefficient and costly when determining tolerance specifications because we cannot afford to include all possible noise effects when using simulation. This chapter describes several simpler, more practical methods, based on the quality loss function, for specifying the tolerances of design parameters for components and products.

3.2 METHODS FOR DECIDING ECONOMICAL SAFETY FACTORS

In this section, economical safety factors (or factors that take into account the financial implications of variances in the objective function), instead of ordinary safety factors, are applied to the specification of tolerances for the objective characteristics of products. In the United States, it is very common to use safety factors, which are usually set at 4, in establishing tolerance specifications. In general, the values of safety factors are determined by engineering experience, taking into account all possible technical problems and malfunctions of a product or technology. In some areas, such as defense and communications, safety factors are even more important and often greater than those of the other areas. However, because the technologies from these two sectors are in wide use in other sectors as well, the tolerance specifications of products in other business sectors are also becoming tighter and tighter. Accordingly, due to changes over time in the way technologies and sectors are related, Japanese Industrial Standards need to be revised every five years. However, outside the defense and communications sectors, the following approach can be used to determine safety factors for specifying tolerance limits that are based on economical considerations instead of engineering experience or judgment.

Let Δ_0 be the functional limit of an objective characteristic of a product (or component). In other words, assuming other objective characteristics are in the normal range (average conditions or nominal values), when the objective characteristic deviates from the target by the amount Δ_0, the product cannot function properly. The functional limit of the objective characteristic Δ_0 is determined by changing this characteristic until the product ceases to function. If one wants to analyze how various noise or outside factors affect the function of a product simultaneously, one may apply Monte Carlo simulation methods or other robust design methods. Standard conditions (or average conditions) for an objective characteristic are defined as the range of values between the upper functional limit and the lower functional limit. The LD50 point is usually used to establish upper and lower functional limits (when an objective characteristic is set at the LD50 point, the product will fail 50% of the time). In general application, one can conduct several experiments to estimate the LD50 points of the upper and lower functional limits of an objective characteristic.

To evaluate economical safety factors, the value of Δ_0 (the functional limit) must first be determined. Consider the ignition voltage of the spark plugs in an automobile engine as an example. Assume that the nominal value of the ignition voltage is 20 kV. (Remember that specifications of the nominal values for system parameters are determined in parameter design, not tolerance design; in parameter design, the nominal values of system parameters are determined to ensure that the objective characteristics of the product are within functional limits. Please refer to Chapter 5 for more details about parameter design.) One can decrease the output ignition voltage to find out the LD50 point for the lower limit, at which point there is only a 50% chance of sparking. Assume that there is only a 50% chance of sparking when the voltage is decreased to 8 kV; therefore, the lower functional limit of the output voltage is -12 kV. Similarly, the voltage is then increased from the nominal value until corona sparking problems occur 50% of the time (when the voltage is too high, corona discharge conditions result, which interfere with proper ignition). This procedure establishes the upper functional limit of the spark voltage. Assume that the upper LD50 point is 38 kV, which translates to an upper functional limit of $+18$ kV. Because the lower and upper functional limits are not the same in absolute value, one might need to use two different economical safety factors for specifying the upper and lower functional limits; therefore, the tolerance specifications for the lower and upper limits of the objective characteristics would also be different. The tolerances for the example of ignition voltage can be specified as $m^{+\Delta_2}_{-\Delta_1}$.

Let the lower and upper functional limits be Δ_{01} and Δ_{02} and the corresponding tolerance specifications be Δ_1 and Δ_2. Also let the corresponding economical safety factors be ϕ_1 and ϕ_2. Then the relationship between the tolerance specifications, the functional limit, and the safety factor can be calculated by dividing each functional limit by the safety factor as follows:

$$\Delta_i = \frac{\Delta_{0i}}{\phi_i} \quad (i = 1, 2) \tag{3.1}$$

In the United States the experiments for estimating the functional limits Δ_0 are usually well planned and conducted. However, engineers in the United States do not usually understand how to apply economical safety factors ϕ_i in tolerance specifications. Since the application of economical safety factors is not well understood, ordinary safety factors (not economical safety factors) for tolerance specifications are usually determined on the basis of engineering judgment or experience. Although some U.S. engineers are beginning to apply probability methods in setting tolerance specifications, these methods are not really appropriate or meaningful in most industrial applications.

For instance, consider the side effects of a prescription drug. Pharmaceutical companies usually use 0.001 of the amount of the LD50 point (the point at which 50% of the patients die of side effects) as the safety factor. However, why use 0.001 as a safety factor? Why not use a different safety factor? What is the special significance of this value? In fact, it is not very meaningful to use this value as a safety factor. Such a statistical method does not take into consideration social and economic factors. This example shows that statisticians do not really understand the exact relationship between the contents of the drug and the corresponding side effects. The other problem here in applying probability or statistical methods is that one does not really know the actual distribution of the side effects. Therefore, it is ridiculous to apply statistical theories to establish safety factors pertaining to the health of many patients.

3.3 LOSS FUNCTION AND ECONOMICAL SAFETY FACTORS

The following method, which is not based on probability or statistical theories, is recommended to calculate the economical safety factors for the tolerance specifications of product development or technology development. The economical safety factor ϕ can be calculated by the following equation:

$$\phi = \sqrt{\frac{\text{Average financial loss when objective characteristic of products exceeds functional limits}}{\text{Average financial loss when objective characteristic just exceeds tolerance specifications of production plans}}} \qquad (3.2)$$

This approach is now beginning to be used in countries outside Japan.

Let the numerator and denominator in Equation (3.2) be designated as A_0 and A, respectively. Then we can rewrite the equation for the economical safety factor as

$$\phi = \sqrt{\frac{A_0}{A}} \qquad (3.3)$$

First, we will explain the reason for applying economical safety factors for tolerance specifications. Let the objective characteristic of a product be y and its target be m, and let the financial loss be $L(y)$ when the objective characteristic y of a shipped product deviates from its target under any kind of uncontrollable noise or operating condition. In addition, assume that the total number of these products in the market is N and the expected operational life of the product is T. Therefore, the financial quality loss of the ith product after being shipped for t years will be $L_i(t, y)$. Generally, $L_i(t, y)$ should be close to zero. However, under some users' operating conditions, some products might not function well; consequently, financial loss will result. The average financial loss for all the products during their whole operating life T will be a good estimator of $L(y)$. Thus, the function, $L(y)$, can be calculated as follows:

$$L(y) = \frac{1}{N} \sum_{i=1}^{N} \int_0^T L_i(t, y) dt \qquad (3.4)$$

In fact, all $L_i(t, y)$'s are independent of each other. However, if N is large enough, $L(y)$ will be close to Equation (3.5) below, which is very similar to the second-order (quadratic) quality loss function described in Chapter 2 [Equation (2.3)]. Therefore, the functional limit Δ_0 and the associated loss A_0 can be used to approximate the loss function as follows:

$$L(y) = \frac{A_0}{\Delta_0^2} (y - m)^2 \qquad (3.5)$$

Since all y's are expected to meet the target m, this is a nominal-is-best type problem, which is discussed at some length in the next section.

In a nominal-is-best type problem, the goal is for the objective characteristic—whether it be the weight, length, or other measurement of a project, for instance—to have a particular, or "nominal," value, and quality loss occurs when the objective characteristic deviates from that value. By contrast, a smaller-is-better type problem has as its goals that an objective characteristic—for example, a particular ingredient in a food—be as small as possible. Conversely in a larger-is-better type problem, a particular objective characteristic would be as great as possible.

Because there is already a target for y in a nominal-is-best type problem, it is easy to estimate the average quality loss of products through this equation. From Equation (3.4), it is obvious that the quality loss in Equation (3.5) is an approximation of the average quality loss caused by uncontrollable noise and operating conditions. Clearly, it is impossible to predict all possible noise and operating conditions under which the shipped products will be used; therefore, it is a good alternative to use Equation (3.5) to estimate the average quality loss caused by all possible noise and operating conditions. Finally, tolerance specifications should be decided based on the trade-off between the average quality loss and the average cost of products.

Of course, the economical safety factor ϕ is a kind of economic parameter (i.e., with financial implications determined by *both* technical and economic means), rather than simply a technical parameter. Therefore, one must first consider the average cost of products when one wants to calculate the economical safety factors for the tolerance specifications of these problems. Let the average cost of products be A when the objective characteristic of products is just outside the tolerance specifications of the production plants. Such products are considered defective. In general, defective products should be adjusted or scrapped. If they are adjusted, A will be equal to the adjustment cost; however, if they are scrapped, A will be equal to the average manufacturing cost of these products. However, these two cases are valid only if the passing ratio (the percentage of products passing inspection and getting shipped) of the products, q, is high (close to 100%). If the passing ratio of the products, q, is below approximately 80%, the cost A must be calculated by the following equation:

$$A = \frac{\text{Average cost of products before being shipped out of plants}}{q} \qquad (3.6)$$

Since the passing ratio of products cannot be directly calculated from the tolerance specifications of production plants, Equation (3.6) is actually a nonlinear function of tolerance specifications. As a result, the value of A cannot be solved directly from this equation. Generally, this equation can be solved by numerical iteration methods.

As an example, let the functional limits of the output voltage of the electrical power supply of a TV set be $\pm 25\%$ of the output voltage. Also, let the average quality loss A_0 of products after being shipped (including adjustment cost and the cost when a TV set does not work) be $300. Assume also that when the voltage of a power supply is just beyond tolerance specifications, one can adjust a resistor in the network so that the output voltage meets its target at a much lower cost, A. Let A, which is equivalent to the sum of the labor cost of adjustment and the cost of a new resistor, be $1. The economical safety factor ϕ can be calculated by the following equation:

$$\phi = \sqrt{\frac{A_0}{A}} = \sqrt{\frac{300}{1}} = 17.3$$

Therefore, the tolerance specifications for this output voltage should be

$$\Delta = \frac{\Delta_0}{\phi} = \frac{25\%}{17.3} = 1.45(\%)$$

Consequently, the tolerance specifications for the output voltage in this example can be specified as follows:

$$115 \pm (115)(0.0145) = 115 \pm 1.7 \text{ (V)}$$

In this example, whereas the functional limits are $115 \pm 25\%$ (or ± 29) V; in comparison, the tolerance specification of the manufacturing plant is only 115 ± 1.7 V. If the output voltage of a power supply is outside of this latter range, the supply should be adjusted before being shipped.

3.4 NOMINAL-IS-BEST TYPE PROBLEMS _____

Remember that in nominal-is-best type problems, quality loss occurs when the objective characteristic deviates from the target. In general, most objective characteristics (e.g., dimensions) of products are nominal-is-best type characteristics.

Consider the case of installing plate glass into window frames. (This example is similar to the case of installing windshields into the front frames of automobiles.) Generally, both window frames and plate glass have their own tolerance specifications. However, to make this example simpler, let us assume that the dimensions of the window frames are very close to their nominal values and their variance is negligible. However, because of the variance in the dimensions of the plate glass, the glass might be too small or too large for the window frames. Let the lower limit of the plate glass be m_1 and the upper limit be m_2. (The objective dimensions could be the horizontal length or vertical length of the plate glass; in this example, use the horizontal length as the objective dimension.) The target value of the dimension m and its functional limits $\pm \Delta_0$ are calculated by the following equations:

$$m = \frac{1}{2}(m_1 + m_2)$$

and

$$\Delta_0 = \frac{1}{2}(m_1 - m_2)$$

Therefore, the tolerance specifications for the dimensions (in this case the horizontal length) of window frames will be $m \pm \Delta_0$. However, the tolerance specifications of plate glass for glass makers (or glass retailers) is not $m \pm \Delta_0$. In fact, the tolerances $\pm \Delta_0$ are the functional limits for customers, not glass makers. Let the average cost when plate glass cannot be inserted into customers' window frames be A_0, which includes the cost of the original plate glass (which cannot be installed into the customers' window frames), the corresponding cutting cost, and the cost of buying another plate glass and its cutting cost. A_0 also includes all transportation costs between the glass maker (or glass retailer) and the customer. Let the average manufacturing cost of plate glass be A. Certainly A_0 will be several times higher than A. The tolerances and economical safety factors of this nominal-is-best type problem can be calculated by Equations (3.1) and (3.3).

Assume that A_0 is \$25 and A is \$3. The economical safety factor can then be calculated by the following equation:

$$\phi = \sqrt{\frac{A_0}{A}} = \sqrt{\frac{25}{3}} = 2.9$$

Now assume that the functional limit for the dimensions of the plate glass is ± 4 mm. One can calculate the tolerance specifications for glass makers by the following equation:

$$\Delta = \frac{\Delta_0}{\phi} = \frac{4}{2.9} = 1.4 \text{ (mm)}$$

If the values of the upper and lower functional limits are different, the upper and lower limit of the tolerance specifications for the glass makers should be calculated separately. However, in many cases the upper and lower limits of tolerance specifications are the same.

3.5 TOLERANCE SPECIFICATIONS FOR SMALLER-IS-BETTER AND LARGER-IS-BETTER TYPE PROBLEMS

The objective characteristics of smaller-is-better type problems are nonnegative and should be as small as possible. In this type of problem, economical safety factors are also calculated using Equation (3.3). For example, let the LD50 point of an ingredient of a prescription drug that might cause side effects be 8000 ppm. The economical safety factors and the tolerance specifications of production facilities for this ingredient can be calculated as below. First, the average cost of the death of one Japanese person can be approximated as follows:

$$A_0 = \text{(Annual GNP of one Japanese) (Average life)}$$

$$= (2 \text{ million } ¥) (77.5 \text{ years}) = (1.55)(10^8 \text{ } ¥)$$

Assume that the unit cost of this drug is 300 ¥. Using Equation (3.3), the economical safety factor for the tolerance specifications for drug manufacturers can be calculated as follows:

$$\phi = \sqrt{\frac{A_0}{A}} = \sqrt{\frac{(1.55)(10^8)}{300}} = 719$$

Therefore, the economical safety factor is 719. Consequently, using Equation (3.1), the tolerance specifications for drug manufacturers for this ingredient can be calculated as follows:

$$\Delta = \frac{\Delta_0}{\phi} = \frac{8000}{719} = 11 \text{ (ppm)}$$

Other examples of smaller-is-better type problems are noises and damaging environmental effects, such as radiation and air pollution.

In larger-is-better type problems, the objective characteristics are also nonnegative but should be as large as possible. The economical safety factor ϕ can also be calculated using Equation (3.3). Let the objective characteristic of a larger-is-better type problem be y and the corresponding loss function be $L(y)$. When y is infinite, $L(y)$ is zero; thus, using Equation (3.4) the average loss function $L(y)$ can be approximated by the following equation:

$$L(y) = \frac{A_0 \Delta_0^2}{y^2}$$

Consider the example of a steel wire. The wire is used to support a load W; thus, its strength is expected to be as large as possible. Assume that the cross-sectional area of this wire is 388 mm², and the unit cost for 1 mm² of this wire is \$4. Therefore, the cost of manufacturing the steel wire is

$$A = \left(\frac{\$4}{\text{mm}^2}\right)\left(388 \text{ mm}^2\right) = \$1552$$

The economical safety factor can then be calculated using Equation (3.3). Assume that A_0, the cost when this wire fails to support the load W, is \$30,000. Therefore, the economical safety factor is

$$\phi = \sqrt{\frac{30,000}{1552}} = 4.4$$

Assume that W is 10,000 lbs. Thus the functional limit of the wire, Δ_0, will be 10,000 lbs. Consequently, the strength specification Δ for plants producing the wire will be 44,000 lbs., which is 4.4 times the functional limit Δ_0.

These methods for using economical safety factors to calculate the tolerance specifications for products can also be used for calculating tolerance limits for components or raw materials. Functional limits of an objective characteristic are usually estimated by its LD50 points (with the objective characteristics of other components or materials assumed to be at the average operating conditions described in Section 3.2).

Let the LD50 point of an objective characteristic be Δ_0 and the corresponding financial loss be A_0. Also let the tolerance specifications and average cost of these components (or raw materials) before being shipped to assembling plants be Δ and A. As indicated above, the economical safety factors ϕ of components or materials of nominal-is-best, smaller-is-better, and larger-is-better type characteristics can all be calculated from Equation (3.3). Therefore, the tolerance specifications Δ for plants manufacturing components can be calculated as follows:

$$\Delta = \frac{\Delta_0}{\phi} \quad \text{for nominal-is-best and smaller-is-better type characteristics}$$

$$\Delta = \phi\Delta_0 \quad \text{for larger-is-better type characteristics}$$

Note that for larger-is-better type characteristics, the effect of the economical safety factor is inverted because the loss is infinite at φ instead of having zero value. If the production process capability is low, one may need to use $A/(1 - p)$ instead of A to calculate the economical safety factor, where p is the defective ratio [see Equation (3.6)].

3.6 METHODS FOR SPECIFYING THE TOLERANCES OF LOWER-LEVEL OBJECTIVE CHARACTERISTICS (UPSTREAM CHARACTERISTICS)

In general, the product planning departments of product assemblers usually determine the tolerance specifications of products. However, the tolerances of components and materials are usually specified at the time of accepting the orders from product assemblers or assembling plants. Therefore, the tolerance specifications for these items are conventionally determined only by the judgment or experience of the product assemblers and component suppliers. The following discussion shows how to specify the tolerances for lower-level objective characteristics (subsystems, components, and materials) that can affect the performance of higher-level objective characteristics (downstream characteristics) of assembled products. In quality engineering, the tolerances for these lower-level characteristics are usually based on parameter design and tolerance design. As discussed in Chapter 5, the nominal values of these lower-level characteristics are determined by parameter design. However, this section explains only how to determine the tolerance limits for lower-level characteristics.

The high-level (downstream) characteristics of shipped products are related to the characteristics of their subsystems. Similarly, the objective characteristics of subsystems are related to the characteristics of their components and constituent materials. Likewise, some factors of manufacturing plants that affect the objective characteristics of shipped products are also the lower-level (upstream) characteristics of these products.

Consider a simple example of a stamped steel product. If the shape of a stamped steel product is not within tolerance specifications, the product must be adjusted. Let the adjustment cost be $A_0 = \$12$. In fact, the thickness and hardness of the steel plate affect the shapes of these stamped products. Assume that the functional limits of the objective characteristic (thickness of the stamped products) are $m \pm 300$ μm. Also assume that if the hardness of the steel plate has a unit change (Rockwell hardness), the thickness of the stamped products will have a deviation of 60 μm from their nominal values. In addition, when a steel plate has a deviation of 1 μm in thickness, the thickness of the stamped products will have a corresponding deviation of 6 μm from their nominal values.

In this example, when the hardness or thickness of the steel plate does not meet specifications, the steel plate is scrapped at a unit cost of $A = \$3$. The tolerance specifications for lower-level characteristics (hardness and thickness of the steel plate) can be calculated from the tolerance specifications of higher-level characteristics (thickness of the stamped products), which are usually determined by loss functions. Let the tolerance specifications of higher-level characteristics be $m_0 \pm \Delta_0$, and the financial loss when these tolerance specifications are not met be A_0. Using Equation (3.5) the quality loss can be calculated as follows:

$$L = \frac{A_0}{\Delta_0^2} (y - m_0)^2 \qquad (3.7)$$

Now, let the lower-level characteristic be x. When x has a unit deviation from its nominal value, the higher-level characteristic y

will have a change of β. Therefore, the right-hand side of Equation (3.7) can be changed into the following form:

$$\frac{A_0}{\Delta_0^2} [\beta(x - m)]^2 \tag{3.8}$$

In this expression, m is the nominal value of the lower-level characteristic x. Let the monetary loss be A when x does not meet the tolerance specifications. Taking Equation (3.8) into the right-hand side of Equation (3.7) and A into the right-hand side of the same equation, we obtain the following equation:

$$A = \frac{A_0}{\Delta_0^2} [\beta(x - m)]^2$$

Therefore, one can determine the value of $\Delta = x - m$ of the lower-level characteristic through the equation above. The tolerance limit for the lower-level characteristic x can be calculated by the following equation:

$$\Delta = \sqrt{\frac{A}{A_0}} \left(\frac{\Delta_0}{\beta} \right) \tag{3.9}$$

The parameters in this equation are as follows:

A_0 = Loss when higher-level characteristics (downstream characteristics) just exceed their tolerance limits

Δ_0 = Tolerances of higher-level characteristics

A = Loss when lower-level characteristics (upsteam characteristics) just exceed their tolerance limits

β = Amount of change in a higher-level characteristic when its lower-level characteristic x has a unit deviation from target value

Both A_0 and A are the financial losses incurred by production plants when the assembled products or components exceed their tolerance specifications and must be adjusted to meet their targets. (Note that neither term includes the cost of inspection and measurement, which will be considered in the next chapter.)

In our example, A_0 is the loss when the stamped steel product is just outside of the higher-level tolerance specifications, $\pm\Delta_0$. Thus we have

$$A_0 = \$12$$
$$A = \$3$$
$$\Delta_0 = 300 \ (\mu m)$$
$$\beta = 60 \ (\mu m)$$

Therefore, using Equation (3.9) the tolerance limits of the lower-level characteristic Δ (the hardness of steel plates) is calculated as follows:

$$\Delta = \sqrt{\frac{3}{12}} \left(\frac{300}{60}\right) = 2.5 \ (H_R)$$

where H_R is Rockwell hardness. Thus the tolerance specifications for the hardness of steel plates will be $m \pm 2.5 \ H_R$. Similarly, the tolerance specifications for the thickness of these steel plates can be calculated using the same equation. Assuming that A_0, A, and Δ_0 all have the same value as above and that β is 6, the calculation yields

$$\Delta = \sqrt{\frac{3}{12}} \left(\frac{300}{6}\right) = 25.0 \ (\mu m)$$

Therefore, the tolerance limits for the thickness of steel plates will be ± 25.0 μm. Of course, this calculation is based on the assumption that when the thickness of a steel plate is very close to its target value, the corresponding steel products will also be very close to their targets for both thickness and shape.

In most cases, lower-level characteristics and higher-level characteristics are linearly proportional to each other. However, in the case of nonlinear relationships, one can employ interpolation methods to establish the relationship between lower- and higher-level characteristics. Figure 3.1 shows a nonlinear relationship between a lower-level characteristic and a higher-level characteristic. However, it is very important that all other lower-level characteristics and environmental conditions be kept at their standard (or average) conditions so that they do not affect this nonlinear relationship.

In Figure 3.1, the tolerance specifications for the lower-level characteristic, Δ_{10} and Δ_{20}, can be determined from the corresponding limits for the higher-level characteristic, $m_0 \pm \Delta_0$. Again, one must be sure that all other lower-level characteristics are kept within standard conditions. Otherwise, they can affect the results obtained when y is shifted.

Note that the upper and lower limits obtained through interpolation might be different. As a result, one might have to calculate these limits separately. Let the lower limit for the lower-level characteristic be Δ_1 and the upper limit be Δ_2. Then using Equation (3.9) where $\beta = 1$, we would calculate their values as follows:

$$\Delta_1 = \sqrt{\frac{A_1}{A_0}}\,(\Delta_{10})$$

$$\Delta_2 = \sqrt{\frac{A_2}{A_0}}\,(\Delta_{20})$$

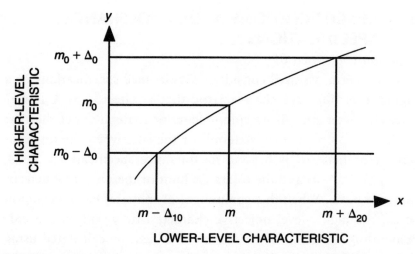

Figure 3.1 Nonlinear relationship between a lower-level characteristic x and a higher-level characteristic y.

In these two equations, A_0 is the cost of the higher-level characteristic when y is just out of its tolerance specifications, and A_1 is the adjustment cost when the lower-level characteristic x is just out of the lower limits of its tolerance specifications (however, materials are usually scrapped if they are not within tolerance specifications). A_2 is the adjustment cost when the lower-level characteristic x is just beyond its upper limit.

Since it is somewhat inconvenient to specify different values for the lower and higher tolerance limits of a lower-level characteristic, one can use the following equation to calculate the tolerance limits for a lower-level characteristic:

$$\Delta = \min(\Delta_1, \Delta_2)$$

Therefore, the tolerances of x would become $m \pm \Delta$.

3.7 MISCONCEPTIONS ABOUT TOLERANCE SPECIFICATIONS

In this section we shall consider the tolerance specifications of a higher-level objective characteristic that is affected by k lower-level characteristics of its components or materials. Let the cost of the ith component or material (or the cost when the ith component or material is beyond its tolerance specifications) be A_i $(i = 1, 2, \ldots, k)$ and the tolerance limit of the ith component or material be Δ_i. When this component or material has a deviation of Δ_i, the higher-level objective characteristic y will have a corresponding deviation of Δ. Therefore Δ^2 can be calculated using Equation (3.7) to obtain

$$\Delta^2 = \left(\frac{A_1 + A_2 + \cdots + A_k}{A_0} \right) \Delta_0^2$$

Therefore, one can compare the sum of the cost of all the components and materials, $A_1 + A_2 + \ldots + A_k$, with A_0, which is the financial loss when the higher-level objective characteristic of the assembled product exceeds its tolerance specifications, to decide whether one should adjust the defective assembled product or scrap it. Three possible cases are discussed below.

Case 1:

$$\frac{A_1 + A_2 + \cdots + A_k}{A_0} \ll 1$$

In this case, if the assembled product exceeds its tolerance specifications, it should be scrapped. The reason is that the average cost of adjusting defective assembled products, A_0, is much more than the sum of the costs of all components and materials. Therefore, it is cheaper to assemble a new product than to adjust a defective one to meet tolerance specifications. If the cost of adjusting defective assembled products is 4 times the sum of the cost of its components, the process capability index of the higher-level characteristics of the assembled product should be about twice those of the components, which can be shown using Equation (3.7). In such a case, no defective product should be shipped out of the plant.

Case 2:

$$\frac{A_1 + A_2 + \cdots + A_k}{A_0} \gg 1$$

In this case, the average adjustment cost for an assembled product is cheaper than the total cost of all components and materials. Therefore, one should adjust the higher-level characteristics of a defective product to meet tolerance specifications, even if the total number of defective assembled products is very large. The reason is that the adjustment cost is much lower than the sum of the cost of all lower-level characteristics; therefore, it is still very reasonable to adjust the defective assembled product rather than to scrap it.

Case 3:

$$\frac{A_1 + A_2 + \cdots + A_k}{A_0} = 1$$

In this case, the adjustment cost A_0 is about the same as the total cost of the components and materials. Since this case does not arise very often, one should check whether the tolerance limits are specified correctly.

In conventional quality engineering, the purpose of specifying tolerance limits for lower-level characteristics is to ensure that the higher-level characteristics of the assembled product will be within the tolerance limits specified by product planning. However, this concept can be misleading. This chapter has demonstrated that one should decide the tolerance specifications for each component on the basis of economics by considering the relationship between the average cost of the ith component of material A_i and the cost of assembled products through Equation (3.9). Of the three cases described above, the first two occur much more commonly than the third.

3.8   INITIAL CHARACTERISTICS AND DETERIORATIVE CHARACTERISTICS

After the initial characteristic has deteriorated over time, it is referred to as the deteriorative characteristic. Let the tolerances of a higher-level characteristic y be specified as $m_0 \pm \Delta_0$. Also let the corresponding cost be A_0 when y does not meet these tolerance specifications. In addition, assume that the objective characteristic y deviates from its nominal value m_0 by an amount β if its lower-level characteristic has a unit deviation from its nominal value m. Finally, let the deteriorative effect of y per year be b and the tolerance limit of b be Δ^*. Using Equation (3.9), the tolerance specification of the initial characteristic of y is calculated as follows:

$$\Delta = \sqrt{\frac{A}{A_0}} \left(\frac{\Delta_0}{|\beta|} \right) \qquad (3.10)$$

where

A_0 = Cost when the initial characteristic of y just exceeds tolerance specifications, $\pm \Delta_0$

Δ_0 = Tolerance limits of the higher-level characteristic

A = Cost when one lower-level characteristic of y just exceeds its tolerance specifications

β = Deviation of y when one lower-level characteristic of y has a unit change

The tolerance specification for a deteriorative characteristic (Δ^*) is calculated using the following equation:

$$\Delta^* = \sqrt{\frac{3A^*}{A_0}\left(\frac{\Delta_0}{|\beta|T}\right)} \qquad (3.11)$$

where A_0, Δ_0, and β are defined as above and

A^* = Cost when the deteriorative characteristic exceeds its tolerance specification

T = Operating life of product in years

Below is an example of how to specify tolerances for the initial characteristic and deteriorative characteristic of a production process. If the illuminance of an integrated circuit (IC) production plant (higher-level characteristic) has a deviation of 50 lux (lx) from the nominal value, production problems can arise, and the average cost for fixing these problems is \$150. Thus,

$$\Delta_0 = 50 \text{ (lx)} \qquad A_0 = \$150$$

However, when the luminosity (lower-level characteristic) of the engineering lamp of this production process has a deviation of 1 candela (cd), the illuminance (higher-level characteristic) will have a corresponding deviation of 0.8 lx. Let the adjustment cost of the luminosity be $A = \$3$. Also, let the cost A^* of replacing an engineering lamp be \$32 when it exceeds its deteriorative tolerance specifications. Thus,

$$\begin{aligned}
\beta &= 0.8 \ (\text{lx/cd}) \\
A &= \$3 \\
A^* &= \$32 \\
T &= 20{,}000 \ (\text{h})
\end{aligned}$$

Consequently, the tolerance limits for the initial characteristic Δ and deteriorative characteristic Δ^* can be calculated using Equations (3.10) and (3.11):

$$\Delta = \sqrt{\frac{A}{A_0}\left(\frac{\Delta_0}{|\beta|}\right)} = \sqrt{\frac{3}{150}\left(\frac{50}{0.8}\right)} = 8.8 \ (\text{cd})$$

$$\Delta^* = \sqrt{\frac{3A^*}{A_0}\left(\frac{\Delta_0}{|\beta|T}\right)} = \sqrt{\frac{(3)(32)}{150}\left(\frac{50}{0.8}\right)\left(\frac{1}{20{,}000}\right)}$$

$$= 0.0025 \ (\text{cd/h})$$

Therefore, the tolerance limits for the initial luminosity are ± 8.8 cd, and the tolerance limit for the deteriorative characteristic of an engineering lamp is 0.0025 cd/h. If the luminosity of an engineering lamp deteriorates significantly, the illuminance of the IC production process will decrease much faster than under normal conditions. Therefore, one needs to calculate the interval between maintenance repairs.

Discussion: About the Problems of Tolerance Specifications

Q: I am doing research on lower-level tolerance specifications when the tolerances of their higher-level characteristics have been specified. In this chapter, you have described how to specify tolerances for each lower-level characteristic individually by considering the relation between the cost of each lower-level characteristic and that of the higher-level characteristic through economical safety factors. However, in my opinion, the objective higher-level characteristic of an assembled product might still be out of its tolerance specifications if all its components (lower-level characteristics) just meet their individual tolerance specifications.

A: Section 3.7 showed that conventional concepts of tightening the variation of lower-level characteristics to make sure that the variation of their higher-level characteristic is within its tolerance specifications are very misleading. For example, let the functional limit (the limit that customers can tolerate) of a quartz watch be ± 10 seconds/month. One should not specify the tolerance specifications for all components of this watch just to make sure that the higher-level characteristic meets these functional limits. Instead, one should adjust the condenser of the watch to ensure that the watch is within tolerance specifications before the watch is shipped. After assembling all components, the tolerance specifications for a watch factory will be only ± 3 seconds/month (actual data), which is much less than ± 10 seconds/month. Also, when a quartz component (which is a lower-level characteristic of the condenser) is out of specifications, the average cost is $A = \$8$. In addition, assume that the cost of adjusting the condenser (which is a higher-level characteristic of a quartz component but still a lower-level characteristic of a watch) is $6. Then the tolerance specifications for the quartz component Δ can be calculated economically from the tolerance specifications for the condenser,

$\Delta_0 = \pm 3$ seconds/month (which is about the same as the tolerance specifications for a shipped watch) by the equation

$$\Delta = \left(\sqrt{\frac{A}{A_0}} \right) \Delta_0 = \left(\sqrt{\frac{8}{6}} \right) 3 = 3.5 \text{ (sec/mo)} \qquad (3.12)$$

Therefore, if the quartz component is within these specifications, the condenser (and thus the watch) will be within the tolerance of ± 3.0 seconds/month.

Q: I think I understand what is meant by the misconceptions about the accumulated variation of higher-level characteristics caused by the variation of lower-level characteristics. I usually receive blueprints from our joint venture companies in the United States. In these blueprints, the higher-level objective characteristics usually exceed their tolerance specifications if all the lower-level characteristics just meet their individual tolerance specifications. I always thought that these tolerance specifications were ridiculous, but now I think I understand them.

A: Yes, you really do understand. One must consider the economic relationship between each component and its higher-level characteristic individually to decide the tolerance specifications for each component. However, it will be explained later how to determine the tolerance specifications for a system that has several lower-level characteristics. In these cases, the tolerance specifications for the components must still be decided through economical safety factors, as calculated by Equation (3.12).

Q: What can I do if I do not have a cheap process for adjusting higher-level characteristics?

A: If there is no cheap process for adjusting higher-level characteristics, the only thing you can do is to treat the product as defective and scrap it. If the adjustment cost of assembled products is much lower than the sum of the costs of all components and materials, one should adjust the higher-level characteristics to meet their tolerance specifications. The methods for specifying the tolerances for lower-level characteristics as described in this chapter are much more meaningful and practical than the conventional methods, which are based on the accumulated variation of higher-level characteristics caused by the variation of lower-level characteristics.

Q: In my opinion, the methods illustrated in this chapter will make the process of specifying tolerances much easier than stacked variation methods, because one can decide the tolerances of each lower-level characteristic directly from A (the cost of the lower-level characteristic) and β (the change in a higher-level characteristic resulting from a unit change in its lower characteristic) without considering the other lower-level characteristics.

A: As a matter of fact, to correctly determine the value β of one lower-level characteristic, one still needs to simulate the effects of several lower-level characteristics simultaneously. This method might be even more reliable than the methods described in this chapter. This approach involves orthogonal arrays to simulate the compounding effects of several lower-level characteristics. However, the methods described in this chapter are practical and reliable enough for general applications.

Exercises

[3.1] The specification for the content of NO_x in car exhaust is below 0.48 g/km according to the (Japanese) 10 mode (1968 regulation). If the content of NO_x is over the tolerance specification before the car is shipped out of the manufacturing facility, the average cost A of adjusting the exhaust system is 6000 ¥. If the exhaust system of a car does not meet the air pollution requirement given, the average cost A_0, of recalling the car is 50,000 ¥.

Calculate the tolerance specifications for the manufacturer for NO$_x$ content.

[3.2] The functional limits (or the range that customers will tolerate) of the output voltage of an electrical network are $\pm 50\%$ of the nominal value of the output voltage. When the output voltage of this network exceeds these limits, the average financial loss is \$180. However, when the manufacturer's tolerance specifications are not met, the average cost of adjusting the network in the plant is only \$0.50. Calculate the tolerance specifications for the manufacturing plant.

[3.3] The target of the output voltage of an electrical network is m (mV), and the tolerance specifications for this voltage are ± 5 mV. If the output voltage of this network exceeds these tolerance specifications, the network is considered defective and is scrapped. The total cost of a defective network is $A = \$0.20$. In addition, when a specific resistor in the network has a deviation of 1 mΩ, the output voltage has a deviation of 0.8 mV. Assume that the nominal value of this resistor has been determined by parameter design and that the nominal value ensures that the output voltage of the network meets the target. However, if the resistance (lower-level characteristic) of the resistor exceeds its tolerance specifications, the average cost is \$0.05. Calculate the tolerance specifications for this resistor.

[3.4]   a. Let the target value of a product dimension (higher-level characteristic) be 300 μm. When this product dimension wears by an amount of 120 μm, the product is out-of-order and the corresponding cost is $A_0 = \$200$. However, this product dimension is affected by a component dimension (lower-level characteristic). Let the tolerance of the initial component dimension and that of the wear rate per year of this component dimension be Δ and Δ^*, respectively. If this component dimension does not meet the tolerance specifications Δ or Δ*, this component is considered defective. Assume that the average cost of a defective component is $0.50. Also assume that when the component dimension has a deviation of 1 μm, the product dimension has a corresponding deviation of 1 μm. The expected operating life of this product is 10 years. Find the tolerance specifications for the wear rate of this component.

b. **Follow-up.** Preventative maintenance of the product is conducted when the amount of wear is 60 μm. In addition, assume that the average wear rate of the product is about 15 μm/year. How many times does preventative maintenance need to be conducted during the operational life of 10 years? The cost of preventative maintenance each time is $30. (This question can be solved by considering the total cost of preventative maintenance and the quality loss caused by the variance of the product dimension.)

4

Quality Management for Production Processes

4.1 OBJECTIVE

Chapter 3 illustrated how to specify tolerances for both higher-level and lower-level objective characteristics. As described in Chapter 2, the limits imposed by manufacturers to control the variance of objective characteristics of products are different from the tolerance specifications for these characteristics. Right now, many American industries use very sophisticated equipment to measure the objective characteristics of their products. However, because some quality managers of these companies do not interrupt the production process as long as their products are within control limits, the objective characteristics of their products are usually uniformly distributed, not normally distributed. In fact, control limits (the adjustment limits used in quality management systems) of products are quite different from the tolerance specifications for these products. This chapter explains how quality management should set control limits for production processes.

4.2 SYSTEM DESIGN FOR THE FEEDBACK CONTROL OF QUALITY MANAGEMENT

When the objective characteristics of products deviate from their targets, one must adjust the production process to bring the objective characteristics closer to their target values. Many types of control systems are used in manufacturing plants to achieve this goal. However, adjusting production processes or operating conditions is usually very costly and time inefficient. Consequently, few companies use feedback quality control systems in their production processes to reduce the variation in the objective characteristics of their products. This chapter describes how to design the optimal feedback system of quality management economically. Below are the definitions of the parameters used in this chapter.

$m \pm \Delta$ = Specifications for objective characteristics of products
A = Cost of one defective product ($)
B = Cost of measurement ($)
C = Cost of adjustment ($)
n_0 = Current measurement interval (units)
u_0 = Current average adjustment interval (units)
D_0 = Current control (adjustment) limit
l = Time lag of current measurement method (units)
n = Optimal measurement interval (units)
D = Optimal control limit
u = Estimated average adjustment interval (units)

In this feedback quality control system, it costs B to measure n units. Thus, the measurement cost per unit of product is

$$\frac{B}{n} = \text{Unit measurement cost} \qquad (4.1)$$

Assume that the average interval between adjustments is u (units) when the control limits (adjustment limits) are $\pm D$. The value of u varies with the value of D and with the stability of the production process. However, the stability of the production process will vary with many factors, such as environmental effects, the wear of tools, and measurement error. Generally speaking, the average adjustment interval u is proportional to the square of the corresponding control limits, D^2. To find the optimal value for D, one must first assign a reasonable initial value to D and then decide the corresponding average adjustment interval u. Next, one checks whether the value of D is appropriate by observing the actual output of the production process. If the value is not appropriate, the value of D must be adjusted iteratively. In addition, the optimal value of u is decided by applying the optimal value of D in the actual production process. The proportional constant between u and D is λ, which is calculated by the following equation:

$$\lambda = \frac{u_0}{D_0^2}$$

At the first stage of designing the quality control system, the initial control limits are usually decided arbitrarily ($D_0 = \Delta/3$ is usually used in Japanese industries, and $D_0 = \Delta$ in American industries, but both assignments are incorrect). One should follow the calibration process mentioned above to find the optimal values of D and u iteratively.

After finding the optimal control limit D, one can calculate the optimal adjustment interval u through the following equation assuming a constant λ:

$$u = u_0 \left(\frac{D^2}{D_0^2} \right) \tag{4.2}$$

The cost of measurement and adjustment can also be taken into consideration in the design of a quality management system by using Equations (4.1) and (4.2). The total measurement plus adjustment cost (using the parameters defined prior to Equation [4.1]) is given as:

$$\frac{B}{n} + \frac{C}{u} = \frac{B}{n} + \frac{D_0^2 C}{u_0 D^2} \tag{4.3}$$

After determining the optimal adjustment level u, one can calculate the actual quality level of the product. This section will explain how to calculate the quality level of products without considering measurement error through the following equation (measurement error will be considered in Section 4.3):

$$\frac{A}{\Delta^2}\left[\frac{D^2}{3} + \left(\frac{n + 1}{2} + l\right)\frac{D^2}{u}\right]$$

In the equation above, the term $D^2/3$ is due to the variation in the objective characteristics of products that are within the control limits $m \pm D$, as illustrated in Figure 4.1. Within the control limits, the objective characteristics are usually uniformly distributed. Therefore, the variation of these uniformly distributed characteristics can be calculated as follows:

$$\frac{1}{2D}\int_{m-D}^{m+D} (y - m)^2 \, dy = \frac{D^2}{3}$$

Assume that the measurement interval for D is n (units). The average number of defective products between a sampled product that passes inspection and another sampled product that is defective will be about $(n + 1)/2$. In the equation, the time lag of the measurement l is also considered. Therefore, the quality loss due to the products whose objective characteristics are over the control limits $\pm D$ can be calculated through the following equation:

$$\left(\frac{n + 1}{2} + l\right)\frac{D^2}{u}$$

Consequently, the total quality loss of the products manufactured between a sampled product that passes inspection and another

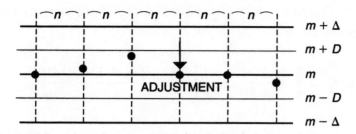

Figure 4.1 Tolerance specifications, $\pm \Delta$, and control limits, $\pm D$.

sampled product that is defective is

$$\left(\frac{n + 1}{2} + l\right)D^2$$

When the average adjustment interval u is applied in this quality control system, the unit quality loss of these products can be calculated by the following equation:

$$\left(\frac{n + 1}{2} + l\right)\frac{D^2}{u} = \left(\frac{n + 1}{2} + l\right)\frac{D_0^2}{u_0}$$

Thus, this unit quality loss in dollars can be expressed by the following equation through the quality loss function:

$$\frac{A}{\Delta^2}\left[\frac{D^2}{3} + \left(\frac{n + 1}{2} + l\right)\frac{D^2}{u}\right] \tag{4.4}$$

The unit quality loss when the measurement and adjustment costs [as in Equation (4.3)] and the quality loss of the products [as in Equation (4.4)] are included can be calculated as follows:

$$L = \frac{B}{n} + \frac{C}{u} + \frac{A}{\Delta^2}\left[\frac{D^2}{3} + \left(\frac{n + 1}{2} + l\right)\frac{D^2}{u}\right] \tag{4.5}$$

Taking Equation (4.2) into Equation (4.5), one obtains the following equation:

$$L = \frac{B}{n} + \frac{D_0^2 C}{u_0 D^2} + \frac{A}{\Delta^2}\left[\frac{D^2}{3} + \left(\frac{n+1}{2} + l\right)\frac{D_0^2}{u_0}\right] \qquad (4.6)$$

If n, u, and D are all replaced by n_0, u_0, and D_0, one gets the following equation:

$$L_0 = \frac{B}{n_0} + \frac{C}{u_0} + \frac{A}{\Delta^2}\left[\frac{D_0^2}{3} + \left(\frac{n_0+1}{2} + l\right)\frac{D_0^2}{u_0}\right]$$

The optimal measurement interval n and the optimal adjustment limit D can be calculated by differentiating Equation (4.6) with respect to n and D and setting the resulting equations equal to zero. With respect to n we get

$$\frac{dL}{dn} = -\frac{B}{n^2} + \frac{A}{\Delta^2}\left(\frac{1}{2}\right)\left(\frac{D_0^2}{u_0}\right) = 0$$

From the equation above, the optimal measurement interval n will be

$$n = \sqrt{\frac{2u_0 B}{A}}\left(\frac{\Delta}{D_0}\right)$$

The optimal adjustment limit D is calculated by the following equation:

$$\frac{dL}{dD} = -2\left(\frac{D_0^2 C}{u_0 D^3}\right) + \frac{A}{\Delta^2}\left(\frac{2}{3}\right)(D) = 0$$

Thus, the optimal adjustment limit D is calculated as follows:

$$D = \left[\left(\frac{3C}{A} \right) \left(\frac{D_0^2}{u_0} \right) \left(\Delta^2 \right) \right]^{1/4}$$

The actual values of n and D are usually rounded to the closest integer values of these optimal values. The unit financial loss for the optimal quality control system L can be calculated through Equation (4.5). However, the term u is estimated through Equation (4.2).

Example

Consider the parameters of the feedback quality control system for a component dimension listed below.

Tolerances of the objective dimension:	$\Delta = \pm 15$ (μm)
Cost of one defective component:	$A = 0.8$
Cost of one measurement:	$B = \$1.50$
Cost of one adjustment:	$C = \$12$
Time lag:	$l = 1$ (unit)
Current measurement interval:	$n_0 = 600$ (units), or 1 measurement every 2 hours
Current control (adjustment) limits:	$\pm D = \pm 5$ (μm)

Current average adjustment
interval: $u_0 = 1200$ (units), or 2
                                             adjustments every
                                             day

(a) Let us evaluate how much the quality is improved by using the optimal control system rather than the current control system. First, using the equations derived above, find the optimal measurement interval n and the corresponding optimal control limit:

$$n = \sqrt{\frac{2u_0 B}{A}}\left(\frac{\Delta}{D_0}\right)$$

$$= \sqrt{\frac{(2)(1200)(1.50)}{0.80}}\left(\frac{15}{5}\right)$$

$$= 201 \text{ (round off to 200 units)}$$

$$D = \left[\left(\frac{3C}{A}\right)\left(\frac{D_0^2}{u_0}\right)(\Delta^2)\right]^{1/4}$$

$$= \left[\left(\frac{(3)(12)}{0.80}\right)\left(\frac{5^2}{1200}\right)(15^2)\right]^{1/4}$$

$$= 3.8 \text{ (round off to 4.0 } \mu\text{m)}$$

Therefore, the unit quality loss for products using these quality control limits can be evaluated by using Equation (4.2) where

$$u = u_0\left(\frac{D^2}{D_0^2}\right) = 1200\left(\frac{4^2}{5^2}\right) = 768 \text{ (units)} \qquad (4.7)$$

If one takes the optimal adjustment interval $u = 768$ units into Equation (4.5), one can get the unit quality loss as follows:

$$L = \frac{B}{n} + \frac{C}{u} + \frac{A}{\Delta^2}\left[\frac{D^2}{3} + \left(\frac{n+1}{2} + l\right)\frac{D^2}{u}\right]$$

$$L = \frac{1.50}{200} + \frac{12}{768} + \frac{0.80}{15^2}\left[\frac{4^2}{3} + \left(\frac{201}{2} + 1\right)\left(\frac{4^2}{768}\right)\right]$$

$$= 0.0075 + 0.0156 + 0.0190 + 0.0075$$

$$= \$0.0496$$

Thus, using a form of Equation (4.6), the current unit quality loss is calculated as follows:

$$L_0 = \frac{B}{n_0} + \frac{C}{u_0} + \frac{A}{\Delta^2}\left[\frac{D_0^2}{3} + \left(\frac{n_0+1}{2} + l\right)\frac{D_0^2}{u_0}\right]$$

$$= \frac{1.50}{600} + \frac{12.00}{1200} + \frac{0.80}{15^2}\left[\frac{5^2}{3} + \left(\frac{601}{2} + 1\right)\left(\frac{5^2}{1200}\right)\right]$$

$$= 0.0025 + 0.0100 + 0.0296 + 0.0223$$

$$= \$0.0644$$

Assume that there are 2000 hours of operation per year. Thus, the improvement in quality due to the optimal quality control system (based on 300 products produced per hour) is

$$(0.0644 - 0.0496)(300)(2000) = 8880 \ (\$/\text{year})$$

(b) Now, estimate the process capability index C_p for the optimal quality control system. The estimated standard deviation of the objective characteristics of these products can be calculated by the equation:

$$\sigma = \sqrt{\frac{D^2}{3} + \left(\frac{n+1}{2} + l\right)\left(\frac{D^2}{u}\right)}$$

From the discussion in Chapter 2, since Δ is 15, the process capability index C_p is calculated as follows:

$$C_p = \frac{\text{Tolerances}}{(6)(\sigma)} = \frac{(2)(15)}{6\sqrt{\dfrac{4^2}{3} + \left(\dfrac{201}{2} + 1\right)\left(\dfrac{4^2}{768}\right)}} = 1.83$$

Therefore, the process capability index is about 1.8. However, recall that measurement error was ignored in this calculation. Accordingly, the process capability may not be quite this good.

4.3 BATCH-TYPE PRODUCTION PROCESSES

A batch-type production process is a procedure in which several units of a product (which constitute one batch) are produced simultaneously. In batch-type production processes, one or more products from each batch is usually sampled and measured to estimate the average objective characteristics for each item in the batch. In addition, one can use these measured values as an estimate of the average objective characteristics of all the products in this production process. However, estimation errors can be introduced if measurement errors of the objective characteristics are not taken into account. [Please refer to the discussion earlier in this chapter for more details about estimation error leading to Equation (4.4).] In the quality loss function for each batch below, the variation of the estimated measurement error σ_m^2 is incorporated:

$$L = \frac{B}{n} + \frac{C}{u} + \frac{A}{\Delta^2}\left[\frac{D^2}{3} + \left(\frac{n+1}{2} + l\right)\frac{D^2}{u} + \sigma_m^2\right]$$

The quality loss due to measurement error is an independent term of the total quality loss per batch. Below is an example to illustrate how to estimate measurement error and the associated quality loss.

Example

Consider an injection molding process at a Japanese manufacturer in which 12 plastic products are produced from each injection shot. In the feedback quality control system, one injected product from every 100 shots is sampled, and the objective dimension of this product is measured and used as an estimate of the average objective dimension of these products. If there is significant deviation from the target, the injection molding process is adjusted.

Assume that there are 100 shots per hour, and the cost for one defective product is 30 ¥. The tolerance for this objective dimension is ± 120 μm, and the current control (adjustment) limits are ± 50 μm. In addition, the adjustment cost of the injection molding process is 1800 ¥, and the average adjustment interval is 800 shots. Finally, the cost of estimating the average objective dimension, B, is 400 ¥ and the time lag is $l = 4$ shots.

(a) Let us determine the optimal measurement interval n; the optimal control (adjustment) limit D; and the magnitude of quality improvement per injection shot and per year at 200,000 shots/year. Since 12 products are produced from each shot, 12 products are treated as one production unit (one batch). (This is very similar to the production process of chemical products, in which 10 tons of chemical materials are processed at the same time and treated as one batch.) Let the total cost be A when all products in one batch are defective. The parameters of this quality control system are defined below:

Tolerances of the objective
dimension: $\qquad \Delta = \pm 120 \ (\mu m)$

Cost of defective products
(per batch): $\qquad A = 30 \ ¥ * 12 = 360 \ ¥$

Cost of one measurement: $\qquad B = 400 \ ¥$

Cost of one adjustment: $\qquad C = 1800 \ ¥$

Time lag: $\qquad l = 4 \ (\text{shots})$

Average current
measurement interval: $\qquad n_0 = 100 \ (\text{shots})$

Current control
(adjustment) limits: $\qquad D_0 = \pm 50 \ (\mu m)$

Current average adjustment
interval: $\qquad u_0 = 800 \ (\text{shots})$

Using equations derived earlier in the chapter, one can calculate the optimal measurement interval n from the parameters above. One can also calculate the corresponding optimal control (adjustment) limit D:

$$n = \left(\sqrt{\frac{2u_0 B}{A}} \right) \left(\frac{\Delta}{D_0} \right)$$

$$= \left(\sqrt{\frac{2(800)(400)}{360}} \right) \left(\frac{120}{50} \right)$$

$$= 101 \ (\text{round off to 100 shots})$$

Therefore, the current measurement interval is already the optimal interval:

$$
D = \left[\left(\frac{3C}{A} \right) \left(\frac{D_0^2}{u_0} \right) (\Delta^2) \right]^{1/4}
$$

$$
= \left[\left(\frac{(3)(1800)}{360} \right) \left(\frac{50^2}{800} \right) (120^2) \right]^{1/4}
$$

$$
= 29 \text{ (round off to 30 } \mu\text{m)}
$$

Consequently, the optimal control limits D are ± 30 μm. It is obvious that these limits are less than the current control limits of ± 50 μm.

The quality loss per batch due to the current method is as follows:

$$
L_0 = \frac{B}{n_0} + \frac{C}{u_0} + \frac{A}{\Delta^2} \left[\frac{D_0^2}{3} + \left(\frac{n_0 + 1}{2} + l \right) \frac{D_0^2}{u_0} \right]
$$

$$
= \frac{400}{100} + \frac{1800}{800} + \frac{360}{120^2} \left[\frac{50^2}{3} + \left(\frac{101}{2} + 4 \right) \left(\frac{50^2}{800} \right) \right]
$$

$$
= 4.00 + 2.25 + 20.83 + 4.26
$$

$$
= 31.34 \text{ ¥}
$$

As a result, the optimal average adjustment interval is

$$
u = u_0 \left(\frac{D^2}{D_0^2} \right) = 800 \left(\frac{30^2}{50^2} \right) = 288 \text{ shots}
$$

Therefore, the quality loss per batch of the optimal control system is

$$L = \frac{400}{100} + \frac{1800}{288} + \frac{360}{120^2}\left[\frac{30^2}{3} + \left(\frac{101}{2} + 4\right)\left(\frac{30^2}{288}\right)\right]$$

$$= 4.00 + 6.25 + 7.50 + 4.26$$

$$= 22.01 \text{ ¥}$$

The gain in quality improvement due to the optimal control system over the current control system is thus

$$(31.34 - 22.01)(100)(2000) = 1,870,000 \text{ ¥/year}$$

(b) Assume that the standard deviation of the measurement error of the current measurement method is 15 μm, while the standard deviation of the error of a new measurement method is only 5 μm. However, if one applies this new measurement method, the cost of one measurement, B, will increase from 400 ¥ to 700 ¥. If we assume that the time lag in both methods is the same, which one is the better choice? First, remember that if the current measurement method is used under the optimal control system, measurement error still exists. Therefore, the equation for quality loss per batch in the optimal control system should include a term that incorporates measurement error, calculated as

$$\left(\frac{A}{\Delta^2}\right)\sigma_m^2 = \left(\frac{360}{120^2}\right)15^2$$

$$= 5.62 \text{ ¥}$$

Therefore, the quality loss per batch under the optimal control system using the current measurement method is

$$22.01 + 5.62 = 27.63 \text{ ¥}$$

On the other hand, the cost of the new measurement method for one measurement is 700 ¥ rather than the current 400 ¥. If the new measurement method is employed, one still needs to calculate the optimal measurement interval n to evaluate the quality loss per batch:

$$n = \left(\sqrt{\frac{2\mu_0 \, B}{A}} \right)\left(\frac{\Delta}{D_0}\right)$$

$$n = \sqrt{\frac{2(800)(700)}{360}} \left(\frac{120}{50}\right)$$

$$= 134 \text{ (round off to 150 shots) or}$$
$$1 \text{ measurement every } 1\frac{1}{2} \text{ hours}$$

The standard deviation of the measurement error of the new measurement method is $\sigma_m = 5$ μm. Therefore, the quality loss per batch is as follows:

$$L = \frac{B}{n} + \frac{C}{u} + \frac{A}{\Delta^2}\left[\frac{D^2}{3} + \left(\frac{n+1}{2} + l\right)\frac{D^2}{u} + \sigma_m^2\right]$$

$$= \frac{700}{150} + \frac{1800}{288} + \frac{360}{120^2}\left[\frac{30^2}{3} + \left(\frac{151}{2} + 4\right)\frac{30^2}{288} + 5^2\right]$$

$$= 4.67 + 6.25 + 7.50 + 6.21 + 0.62$$

$$= 25.25 \text{ ¥}$$

Because the current measurement method is less accurate than the new measurement method, the estimated savings (gain) due to the new measurement method will be

$$(27.63 - 25.25)(100)(2000) = 476,000 \text{ ¥/year}$$

Clearly, it would be appropriate to shift to the new measurement system.

Discussion: Prediction and Adjustment

Q: In a production process, the objective characteristic of every product is measured. Assume that the objective characteristic of the first product, y_0, is a little bit larger (for example, 8 μm) than the target value m. One does not need to adjust the process at once. Assume that after 2 hours, 300 products are produced. Thus, the average objective characteristic for these 300 products can be estimated by the following:

$$\bar{y} = \frac{y_1 + y_2 + \cdots + y_{300}}{300}$$

One can compare this average objective characteristic, $\bar{y}$, with y_0 to adjust the production process to meet the target. Is this a kind of feedback control system?

A: Yes. After measuring y_0 (the objective characteristic of the first product) and $\bar{y}$, one can estimate the prediction error, σ_p, by the difference between y_0 and $\bar{y}$. In addition to this prediction error, one can use a damping factor β to adjust the process to reduce the variation of the objective characteristics from their target values.

Let the target of an objective characteristic be m. If one wants to adjust the production process from the first measured objective characteristic y_0, one can apply the amount $-\beta(y_0 - m)$ instead of the full deviation $-(y_0 - m)$ to adjust the process. The value of β is defined as follows:

$$\beta = \begin{cases} 0 & \text{when } F_0 = \dfrac{(y_0 - m)^2}{\sigma_p^2} \leq 1 \\ 1 - \dfrac{1}{F_0} & \text{when } F_0 > 1 \end{cases}$$

However, this method is still not very convenient. If a computer is not available, one may use the following method for approximating the value of β:

$$\beta = \begin{cases} 0 & \text{when } F_0 \leq 1.5 \\ 1 & \text{when } F_0 > 1.5 \end{cases}$$

The factor β should have a value of 1 when almost all the measured data of the objective characteristic are larger than m (or all smaller than m).

Exercises

[4.1] Let the tolerance specifications for the diameter of one component be $m \pm 40$ μm and the cost for one defective component, A, be \$1.20. The production rate is 120 units per hour. The current measurement interval is one measurement of one sample per hour, and the control limits are ± 10 μm. In addition, the measurement cost of one sample, B, is \$1.20, and the time lag is 2 units per measurement. The adjustment cost C is \$3 when the diameter of a measured component exceeds the control limits. Finally, the average adjustment interval is 20 hours.

a. Design the optimal feedback control system for this production process. What is the savings from using the optimal control system compared with the current control system? Assume that there are 2000 hours of operation per year.

b. Calculate the process capability indexes C_p for the current and the optimal quality control systems.

c. Assume the time needed for one measurement is 2 minutes, and the time for one adjustment is 8 minutes. What are the actual man-hours (operation hours not including the time of measurement and adjustment) required for this production process? *Hint*: man-hours can be calculated by the following equation:

Man-hours = (Total operation hours) − [(Number of measurements per year)(2 minutes/60) + (Number of adjustments per year)(8 minutes/60)]

[4.2] In a coating process, 800 products are coated simultaneously and treated as one batch. The tolerance specifications for the coating film are ±3.5 μm. The loss of one unit is $A = \$3$. The average measurement cost for each batch is estimated to be $15. Each batch is sampled and measured, and the control limits are ±0.5 μm from the target value. The time lag l is zero, and the adjustment cost C is $30. In addition, the average adjustment interval is 5 batches. The production rate is 6 batches per day, and there are 250 working days per year.

a. Design the optimal feedback control system for this coating process. What is the quality improvement due to the optimal control system compared with the current control system?

b. The standard deviation of the error due to the current measurement method is $\sigma_m = 0.3$ μm. Assume that a new measurement method has a standard deviation of the error of $\sigma_m = 0.15$ μm; however, this new method costs $(1.5)(15) = \$22.50$ per batch. The time lag is assumed to be zero. What are the savings due to the new measurement method?

5

Parameter Design

5.1 ABOUT PARAMETER DESIGN _____

The responsibilities of technology departments in developing new products are listed below:

1. **Product planning.** Deciding the functions, prices, and operating lives of new products.

2. **Product design.**
   a. *System choice.* Choosing the most appropriate system for product development from all possible systems that can perform the objective functions.
   b. *Parameter design.* Deciding the optimal nominal values for the parameters of the chosen systems.
   c. *Tolerance design.* Finding the optimal trade-off between quality loss due to the variation of the objective functions and the cost of high-grade components.

3. **Production process design.**
   a. *System choice.* Choosing appropriate production processes.
   b. *Parameter design.* Deciding the optimal nominal values for the parameters of these production processes.
   c. *Tolerance design.* Deciding the tolerance specifications for the parameters of these production processes.

To ensure that all these tasks are performed smoothly and efficiently, all the technologies needed to design and manufacture new products must be completely developed even before new products are planned. In other words, new products cannot be developed smoothly and efficiently if the technologies needed for new product development and production are not available. The only way for high-technology industries to survive in the highly competitive global market is to develop robust technologies before new products are planned. Component suppliers should likewise be utilizing robust technologies to keep apace of new product development. If all the technologies are in place before new products are planned, one only needs to make very simple adjustments in the actual production process in order to begin mass production of new products.

There are three important factors in research related to robust technology development:

1. technological readiness
2. flexibility
3. reproducibility

Technological Readiness

A technology is considered ready when all necessary information for the functional design of a new product has been provided prior to product planning. In addition, design engineers for new products should completely understand the capabilities and details of newly developed production technologies. In this way only simple adjustments will be required to put new products into mass production, instead of having to design a new production process. As a result, robust technology development should constitute the upstream research activities of new product planning and product design. However, one can see that it is an enormous challenge to design appropriate technologies before new products are developed.

Flexibility

If new technologies are developed for only one type of product or one type of production process, the development efficiency for all possible new products will be low. Therefore, any newly developed technologies should be applicable to the development of many new types of products. To ensure such flexibility, one needs to conduct research to maximize the robustness (some companies may call it "stability") of newly developed technologies and to determine all possible adjustment or tuning factors so that these technologies can be modified to suit actual production processes. In quality engineering, many types of signal-to-noise (S/N) ratios are usually calculated to measure the robustness for the functions of products, processes, or technologies. In this chapter we discuss the appropriate S/N ratios for gauging the robustness of new technologies and the techniques for maximizing the S/N ratios through parameter design.

Reproducibility

Robust technology development ensures that the design for production processes is the best for actual production conditions and that the design of new products is the best for customers' operating conditions. As mentioned above, technological readiness is a very important factor for robust technology development. Thus, technology development should be conducted before planning new products so that products can be developed more efficiently. However, in most enterprises the research activities of technology development are conducted in research departments. (Many companies call these research activities "basic technology research.") Consequently, the technologies developed usually cannot be applied to actual mass production conditions smoothly and efficiently. The reason is that the operating conditions in research departments are quite different from the actual production processes or actual users' operating conditions. In general, technologies that are not robust under actual production or operating conditions are useless and wasteful.

Research done in research departments should be applied to actual mass production processes and should take into account all kinds of operating conditions. Therefore, research departments should focus on technology that is related to the functional robustness of processes and products, and not just to meeting customers' requirements. Bell Laboratories called this type of research "two-step design methods." In two-step design methods, the functional robustness of products and processes is first maximized and then these products and processes are tuned to meet customers' requirements.

As illustrated above, measuring robustness by S/N ratios is very important to technological readiness, flexibility, and reproducibility at the stage of technology development. Using S/N ratios as a measurement of technology originated in the communications industry. However, after several decades of development, S/N ratios are now used in all fields of engineering. In quality engineering, maximizing S/N ratios by calibrating system parameters is called parameter design. However, many companies in the West call these methods *Taguchi methods* to distinguish them from other parameter design methods.

Measuring the functional robustness of products or technologies by calculating appropriate S/N ratios is a very important task in quality engineering. In addition, defining the input signals and the ideal function of the product or process, choosing the objective characteristics, selecting levels for the noise factors, and choosing the tuning processes for mass production are all part of the process of selecting appropriate S/N ratios.

The best way to convey the basic concepts of robust technology development is by presenting the simple, easily understandable examples provided below. Two examples pertain to robust product design (the development of resistors and paper-feeding mechanisms), and two pertain to robust production processes (welding and injection molding). The basic ideas expressed in these examples can be applied to other areas of engineering.

5.2 FUNCTION VERSUS QUALITY

In the research conducted by the makers of electrical resistors, resistance values are measured to improve the quality of resistors. This is similar to spring makers measuring spring constants to improve the quality of their springs. However, resistance values and spring constants are measures of quality, not objective functions of the products. In fact, designers apply the *function* of resistors, not the resistance values of resistors, when they design electrical networks. Resistors convert current into voltage (or voltage into current). Thus, the input signal is voltage (or current). In the example below, the input is voltage. The ideal function between voltage M and current y under actual operating conditions is shown by the equation

$$y = \beta M \tag{5.1}$$

However, under actual operating conditions, the current y is a much more complicated function than Equation (5.1) indicates, involving factors such as input voltage, alternative frequency, temperature, and humidity. Let these factors be $x_1, x_2, \ldots, x_n$. Thus, one obtains the following equation:

$$y = f(M, x_1, x_2, \ldots, x_n) \tag{5.2}$$

If the value of the right-hand side of Equation (5.2) is very close to that of Equation (5.1), the function of this resistor is robustly designed. The right-hand side of Equation (5.2) can be separated into the sum of an ideal function, which is the same as that in Equation (5.1) and a quantity that is a deviation from the ideal function:

$$y = f(M, x_1, x_2, \ldots, x_n)$$
$$= \beta M + [f(M, x_1, x_2, \ldots, x_n) - \beta M]$$

The linearly proportional term βM is called the useful part; the part in brackets, which is the deviation from the useful part, is called the harmful part because it is harmful to the robustness of y. In communications engineering, engineers usually use the following S/N ratio (η) to calculate the ratio between the useful part and the harmful part:

$$\eta = \frac{\left(\dfrac{\partial f}{\partial M}\right)^2}{\left(\dfrac{\partial f}{\partial x_1}\right)^2 \sigma_{x_1}^2 + \cdots + \left(\dfrac{\partial f}{\partial x_n}\right)^2 \sigma_{x_n}^2} \tag{5.3}$$

where η is the S/N ratio and σ is the observed standard deviation for each factor.

This S/N ratio is a measure of robustness; the higher the ratio, the less harm variations cause to the system. The S/N ratio in the equation above contains the part of the deviation from the useful part over all ranges of the input signals. The deviation from the useful part is the major error of this function. However, although this type of S/N ratio shows the ratio between the useful and harmful parts, it cannot evaluate the total variation of the harmful part. This is the main disadvantage to this type of S/N ratio. In addition, computing the values of the numerator and denominator of the right-hand side of Equation (5.3) is quite laborious. Therefore, a more efficient way to express the relation between the useful part and the harmful parts would be quite helpful. One example is the S/N ratio described in the next section, which can be used in the design and development of resistors.

5.3 IDEAL FUNCTION AND SIGNAL-TO-NOISE RATIO _____

In this section we consider electrical resistors used in a type of electrical network to illustrate the basic concepts of robust technology development for product design. As mentioned in Section 5.1, it is almost impossible to adjust the qualities of electrical networks at the stage of technology development; instead, one should adjust the new product to meet customers' requirements at a later stage. At the stage of technology development, one should improve the robustness of subsystems and components, such as resistors. Since the resistance value of resistors is not really the function of resistors, measuring resistance values is not a very efficient way to improve the technology of resistors; instead, one needs to measure the functions of resistors. As stated in the last section, the function of resistors in electrical networks is to convert voltage into current.

Assume that the range of the input voltage can be classified into k levels, $M_1, M_2, \ldots, M_k$. In addition to leveling the range of input voltage, all possible noise factors are also grouped together into a compound noise factor N. All possible frequencies, environmental conditions, voltages, and deterioration effects should be included in this compound noise factor. Assume that the factors

TABLE 5.1 Input and output data for electrical resistors

	$M_1 \, M_2 \cdots M_k$	L (linear proportional term)
N_1	$y_{11} \, y_{12} \cdots y_{1k}$	$L_1 = M_1 y_{11} + \cdots + M_k y_{1k}$
N_2	$y_{21} \, y_{22} \cdots y_{2k}$	$L_2 = M_1 y_{21} + \cdots + M_k y_{2k}$

comprising N can be separated into two levels, N_1 and N_2, as follows:

N_1: Noise factors that can slightly shift the output current from the linear proportional term

N_2: Noise factors that can significantly shift the output current from the linear proportional term

The value of the output current y can be measured by the experimental layout of k levels of voltage and two levels of compound noise factors. The resulting data will be the left half of Table 5.1. Each y is the output current in its range of input voltage for appropriate noise factor. Remember that the ideal function between the input voltage M and the output current y is shown as the linear term in Equation (5.1). However, because of the noise effects mentioned above, the actual function was changed to the more complicated function, as shown in Equation (5.2). Remember also that Equation (5.2) was divided into two parts, the ideal function (useful part) and the deviation from the ideal function (harmful part), as follows:

$$\overbrace{}^{\text{Useful part}} \qquad \overbrace{}^{\text{Harmful part}}$$

$$y = \beta M \quad + \quad [f(M, x_1, x_2, \ldots, x_n) - \beta M]$$

The relative magnitude of the useful part can be evaluated by the square of the linear proportional coefficient β. The relative magnitude of the harmful part can be evaluated by the variance σ^2 (mean square error, or the second-order error). One can evaluate the S/N ratio, β^2/σ^2, from the data in Table 5.1 through the following equation:

$$\eta = 10 \log \frac{1}{r}\left(\frac{S_\beta - V_e}{V_N}\right) \tag{5.4}$$

The variables in this equation are calculated by the following equations, where DOF is the degrees of freedom:

$$r = 2(M_1^2 + M_2^2 + \cdots + M_k^2) \tag{5.5}$$

$$S_\beta = \frac{(L_1 + L_2)^2}{r} \qquad (DOF = 1)$$

$$V_e = \frac{S_e}{2k - 2}$$

$$V_N = \frac{S_{N*\beta} + S_e}{2k - 1}$$

where

$$L_1 = M_1 y_{11} + M_2 y_{12} + \cdots + M_k y_{1k} \tag{5.6}$$

$$L_2 = M_1 y_{21} + M_2 y_{22} + \cdots + M_k y_{2k} \tag{5.7}$$

$$S_T = y_{11}^2 + y_{12}^2 + \cdots + y_{2k}^2 \qquad (DOF = 2k)$$

$$S_{N*\beta} = \frac{(L_1 - L_2)^2}{r} \qquad (DOF = 1)$$

$$S_e = S_T - S_\beta - S_{N*\beta} \qquad (DOF = 2k - 2)$$

V_e is the mean square error of nonlinearity, V_N is an error term of nonlinearity and linearity, and L_1 and L_2 are the weighted linear proportional terms from which the slopes of N_1 and N_2 deviate (see the right side of Table 5.1). S_T is the magnitude of the total sum of squares of the output current y_{ij}. S_β is the sum of squares of the useful part and S_e is the sum of squares of the error terms. $S_{N*\beta}$ is the measurement of the difference between N_1 and N_2. The term r is a measurement of the magnitude of the input signals. Using the dynamic-type S/N ratio above to measure the functional robustness of resistors will make the technology development of resistors much more efficient.

In addition to the dynamic-type S/N ratio, one can use nominal-is-best type S/N ratios to evaluate the functional robustness of resistors at various levels of input current. Here the level is fixed and the mean is adjusted on target. However, this latter type of S/N ratio is not very efficient when the resistance values are very small, because it can lead to unreliable conclusions. After determining the optimal setting for controllable factors, one can adjust the resistance values of the resistors to meet various targets. It is the duty of production departments to perform such adjustments using process control methods.

It is completely feasible for technology departments to develop new technologies before new products are planned. As mentioned early in this chapter, technology readiness is very important in developing new products efficiently. To achieve this goal, one can follow the two-step design methods described above. Because maximizing robustness is separate from the tuning (adjustment) process, two-step design methods are quite different from the response surface optimization methods (one-step optimization) employed in conventional statistical procedures. If one uses a one-step optimization method, one might not be able to determine the optimal design easily because of significant interactions among the controllable factors. Another disadvantage of one-step optimization methods is that the target values for new products are usually not specified at the stage of technology development. If one specifies a single target and then uses a one-step optimization method to determine the optimal setting for the controllable factors, one will lose flexibility and technological readiness. Because one-step optimization methods are useful for developing only one product and not various products, they are not efficient for technology development.

5.4 A DYNAMIC-TYPE PROBLEM: INJECTION MOLDING _____

For systems of products, production processes, or components that have multiple inputs and outputs, one should use dynamic-type characteristics (S/N ratios) as objective measurements of functional robustness. In this type of problem, the functional robustness can be tested by comparing input signals with the corresponding functional outputs:

The function of injection molding is for injected products to assume specified shapes and dimensions. Thus, the dimensions of injected products are functions of the dimensions of the molds. Therefore, the input signals are the dimensions of the molds, and the outputs are the dimensions of the injected products. The plastic resin (or ceramic powder) is only the raw material of the injection molding process and not really the input signal. Thus the only input signals are the dimensions of the injection molds.

The objective of robust technology development for dynamic-type problems such as injection molding is to make sure that the various outputs meet their corresponding targets. The ideal relationship between input signals and output signals is illustrated by the relationship between M and y in Equation (5.1). It is very important to make the relationship between M and y close to this ideal function at the stage of robust technology development. In general, this ideal relationship is linear (but not always). In the case of injection molding, let the M's be the dimensions of the molds and the y's the dimensions of the injected products.

In actual injection molding, because of various factors such as the position of the injection gates, environmental effects, the varying plasticity of raw materials, and uneven temperature distribution in injection molds, the M's and y's are usually not linearly proportional to each other. In fact, their relationship is usually a very complicated function, as shown in Equation (5.2). In this function, let $x_1, x_2, \ldots, x_n$ represent the factors of the injection molding process.

Since this is not scientific research, it is not really important to find out the exact function of Equation (5.2). The important task is to determine the optimal settings for the controllable factors of the injection molding process so that Equation (5.2) is close to Equation (5.1). In quality engineering, these methods are also called "parameter design methods." The objective of these methods in this example is to improve the functional robustness of the injection molding process in converting the dimensions of injection molds into the dimensions of products. The process will be "functionally robust" if it produces the desired part (i.e. satisfies the design intent) for a wide range of part dimensions. Below are the steps necessary to conduct robust technology development for the injection molding process.

Step 1. Find the signal factors (input signals) and the uncontrollable error factors (noise factors) of the molding process and their ranges. Then determine appropriate levels for these factors.

In our example, one must first decide the ranges (for example, from 1 mm to 100 mm) of the dimensions of the injection molds (input signals). One can certainly make some experimental molds where the dimensions are within specified ranges. However, one can also use existing molds that are within these ranges as the experimental molds. Among all possible noise factors, the location of the injection molding machine and the order of the injection shots are especially important. Next, one chooses k dimensions for one mold, $M_1, M_2, \ldots, M_k$. Now let the first injection shot be N_1 and the tenth shot (for example) be N_2. Collect the data using Table 5.2.

In this table, the dimensions of the injected products for N_1 and N_2 are $y_{11}, \ldots, y_{1k}$ and $y_{21}, \ldots, y_{2k}$.

Step 2. Choose as many controllable factors as possible, select levels for these factors, and assign these levels to appropriate orthogonal arrays.

Controllable factors are factors that can be adjusted to different levels to improve the functional robustness of the injection molding process. The levels of the following controllable factors can be freely decided by engineers:

▶ Types of plastic resin and the amount or types of additives
▶ Injection pressure or speed
▶ Preheating temperature of injection molds
▶ Temperature of plastic resin
▶ Holding time
▶ Cooling method

One should choose appropriate levels for these controllable factors and assign them to appropriate orthogonal arrays. Continuous

TABLE 5.2 Input and output data for the injection molding process

	$M_1\ M_2 \cdots M_k$
N_1	$y_{11}\ y_{12} \cdots y_{1k}$
N_2	$y_{21}\ y_{22} \cdots y_{2k}$

quantitative factors such as temperature and pressure are usually assigned to three levels. One can then conduct experiments to measure the dimensions of injected products as in Table 5.2.

Step 3. Calculate S/N ratios from the experimental data.

One can calculate the S/N ratio as in the previous example using Equation (5.4):

$$\eta = 10 \log \frac{1}{r}\left(\frac{S_\beta - V_e}{V_N}\right)$$

where all terms here are defined as in Section 5.3.

This S/N ratio (η) is the measurement of the functional robustness of the converting technologies (or shaping technologies) of the injection molding process.

Step 4. Determine the optimal conditions for the injection process and estimate the proportional constant β between mold dimensions and product dimensions.

To find the main effect of each level of the controllable factors, one can calculate the average value of the S/N ratio for each level. One can then choose the levels with the maximum average S/N ratios as the optimal settings for these controllable factors. These optimal settings will be the optimal conditions for the injection molding technology between the mold dimensions and the product dimensions. To compare the initial conditions and the optimal conditions, one can conduct verification experiments under the two compound noise conditions N_1 and N_2 and calculate the S/N ratios for these two conditions. If the S/N ratio of the optimal setting is much larger than that of the initial setting, this optimal setting will be very robust and reliable. In addition to the optimal conditions, one also needs to calculate the shrinkage ratio β between the mold dimensions and the product dimensions by the following equation:

$$\beta = \frac{L_1 + L_2}{r}$$

where r, L_1, and L_2 are from Equations (5.5), (5.6), and (5.7), respectively.

Step 5. Design new molds based on the S/N ratio and the estimated shrinkage ratio β. If the S/N ratio of the optimal setting is not much larger than that of the initial conditions, one still needs to adjust the dimensions of the injection molds for the new products. However, if the S/N ratio is very large, adjusting the dimensions of the molds is unnecessary, and the new molds can be placed directly into mass production.

5.5 A DIGITAL-TYPE PROBLEM: AUTOMATED SOLDERING _____

Several years ago the production technology research department of the Western Electric Company sought help in developing a new technology for automatic soldering. Similar technology was also being developed by a Japanese company. To determine the physical problems in production facilities, this Japanese company chose the boards used on the production line for its research. Then all possible soldering conditions that might arise on the production line were re-created, and the automated soldering process was adjusted to reduce the ratio of defective soldered boards to a minimum. After solving all the problems for one type of board, engineers applied the newly developed soldering automation process to other types of boards and checked whether new problems occurred. If so, engineers adjusted the new process to solve these problems.

The research activities described above are used to solve the physical problems in current production plants, but not to develop new technology. This type of research refines current production methods quickly by adjusting existing production systems and conditions. In contrast, technology development focuses on the research and development of a completely new generation of technology. In the United States, research activities that are devoted to the refinement and improvement of current production methods are not considered technology development.

In Japan, engineers like those in the example above usually conduct research on improving existing production conditions to meet the quality requirements specified by new product planning. However, when a new generation of products is planned and put into mass production, many new problems can arise in existing production processes. The number of problems due to a new generation of products is usually very great and cannot be anticipated. As a result, it is usually necessary to do research to improve quality every time a new generation of products appears. This approach is typically very inefficient. This inefficiency generates longer product development cycles and greater product development cost. Japanese companies tend to confine their research to improving the quality of current products and current production processes and systems, rather than on developing new generations of technology.

The following example shows how to develop new technologies after selecting a new process such as the automated soldering process mentioned above. Since the experimental results of the soldering process are not quantitative values, one should use operating window methods to evaluate the quality of the soldering process. Next, one should consider all possible noise conditions, such as the variation in heat needed for the soldering terminals, the density of various solders, the structure of the boards, and the amount of dirt on the soldering terminals. Taking all these noise conditions into consideration, one gets a compound noise factor N. Three levels of N are as follows:

N_1 = Soldering terminals that need the least heat energy

N_2 = Standard soldering terminals

N_3 = Soldering terminals that need the most heat energy

One experimental board that is soldered by three pairs of terminals, N_1, N_2, and N_3, is shown in Figure 5.1. Then one should prepare enough experimental boards (as in the figure) to determine the threshold values for the heat energy needed for the following two conditions:

x = Threshold value of the maximum soldering power for nonsoldering conditions

y = Threshold value of the minimum soldering power for bridging conditions

(Under these conditions, the soldering terminals overheat and bridging problems occur.)

Next, one should conduct experiments to find out these two threshold values for the three noise conditions. The power applied to each pair of soldering terminals must be carefully observed to estimate the LD50 points of the threshold values. Let the corresponding threshold values for the soldering terminals N_1, N_2, and

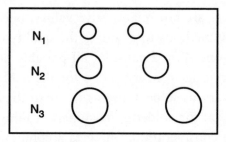

Figure 5.1 Experimental board soldered by three pairs of soldering terminals

N_3 be x_1, x_2, x_3, y_1, y_2, and y_3. The purpose of this experiment is to determine the optimal soldering conditions that maximize the S/N ratio. This S/N ratio can be evaluated by the following equation:

$$\eta = -10 \log\left[\left(\frac{x_1^2 + x_2^2 + x_3^2}{3} \right)\left(\frac{1}{3} \right)\left(\frac{1}{y_1^2} + \frac{1}{y_2^2} + \frac{1}{y_3^2} \right) \right]$$

If the S/N ratio becomes large after tuning the controllable factors, the soldering process will be very robust and reliable. In this way the technology for the soldering process can be developed, and many types of boards can be well soldered by this technology. (A similar example is the development of a paper-feeding mechanism for copying machines, conducted by the Fuji-Xerox Corporation. This will be discussed in the next section.)

The purpose of this type of experiment is to find the optimal settings for the controllable factors and production conditions to extend the ranges of the operating window under various soldering conditions. In developing the automated soldering process, engineers might be able to get a good deal of new technology information, such as the highest density of electrical wires and the widest permissible range for the heat capacity of the components.

Comparing the two types of development processes, process refinement and technology development, we can see that the activities of research institutes in the western industrialized countries are quite different from those in Japanese companies, which focus on research in product development. However, in the West, the production plants are responsible for maintenance only. They are free only to tighten the tolerances of the products under various production conditions. The Western Electric Company in Richmond was once urged to conduct several physical experiments in its production plants, but the suggestion was rejected by a quality assurance agent from Bell Laboratories. Since the quality assurance agent was not really familiar with off-line quality control, only concepts relating to on-line quality control were considered. In western industrialized countries, the research activities performed by technology development are usually conducted in laboratories. Therefore, most research activities focus only on the technological aspects of the design of products and production processes, rather than on how to develop robust products or robust production processes.

5.6 ANOTHER DIGITAL-TYPE PROBLEM: A PAPER-FEEDING MECHANISM_____

The function of a paper-feeding mechanism in a copying machine is to feed exactly one piece of paper each time the mechanism receives an input signal. When the mechanism does not feed any paper, it is called "misfeeding." When two or more pieces of paper are fed into the copy machine at the same time, it is called "multifeeding." As shown in Figure 5.2, this mechanism applies friction between the feed roller and the paper, and the torque of the feed roller feeds the paper into the copying machine. The friction force

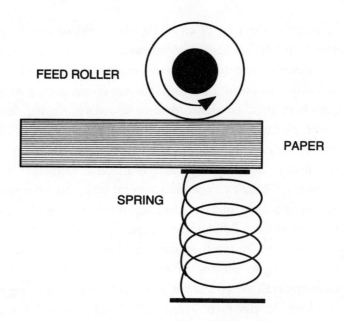

Figure 5.2 Paper-feeding mechanism

between the feed roller and the paper is determined by the force of the spring applied below the paper tray. When the spring force is zero, no paper will be sent out of the paper tray (misfeeding). When the spring force is properly set, one piece of paper will be sent out. Let the threshold value of the spring force for sending one piece of paper be x (gram-force). When the spring force becomes too large, two or more pieces of paper might be sent out of the tray (multifeeding). Let the threshold value of the spring force for multifeeding conditions be y (gram-force). Finally, let the three levels of a compound noise factor N be defined as follows:

N_1 = Conditions when multifeeding arises frequently (for example, lightweight paper when static electricity is high)

N_2 = Standard conditions

TABLE 5.3 Operating window of threshold spring forces

	Threshold value of misfeeding	Threshold value of multifeeding
N_1	x_1	y_1
N_2	x_2	y_2
N_3	x_3	y_3

N_3 = Conditions when misfeeding arises frequently (for example, where the paper has low friction force with the roller because the roller is light)

Under these three noise conditions, one can vary the spring force to determine the threshold values for misfeeding and multifeeding, as shown in Table 5.3. However, notice that the conditions $x_i = y_i$ ($i = 1, 2, 3$) can occur. Under such conditions the range of the operating window is zero. After collecting the data in Table 5.3, one can calculate the S/N ratios for the operating window as follows:

$$\eta = \eta_x + \eta_y \qquad (5.8)$$

where

$$\eta_x = -10 \log\left(\frac{x_1^2 + x_2^2 + x_3^2}{3}\right) \qquad (5.9)$$

$$\eta_y = -10 \log\left[\frac{1}{3}\left(\frac{1}{y_1^2} + \frac{1}{y_2^2} + \frac{1}{y_3^2}\right)\right] \qquad (5.10)$$

The three threshold values x_1, x_2, and x_3 are smaller-is-better type characteristics; thus, they should be as small as possible. In contrast, the three threshold values y_1, y_2, and y_3 are larger-is-better type characteristics and so should be as large as possible. Thus, these two S/N ratios can be evaluated from Equations (5.9) and (5.10) individually. The sum of Equations (5.9) and (5.10) can be used to calculate the range of the operating window of the paper-feeding mechanism. This sum [Equation (5.8)] will be the measurement of the functional robustness of this mechanism. This S/N ratio is a relative measurement value for one single system; thus, it is not very meaningful to compare the S/N ratio of the operating window of one system with that of another.

5.7 CONCLUSION

The only way for companies to survive in the present era of rapidly changing technology is to make the change from product-oriented research to technology-oriented research. To do so, technological research instead of product design should be done before planning new products. In addition, technology departments should offer enough technological information to product development and product design departments before new products are planned. There are two important steps in this type of research:

1. improving the functional robustness of technologies by maximizing S/N ratios and then choosing appropriate tuning factors, and
2. adjusting the technologies by applying the tuning factors to meet various targets or customers' requirements at the stage of product planning or product design.

Only S/N ratios efficiently measure the functional robustness of new technologies at the stage of technology development. Tuning factors should be chosen from the controllable factors that do not affect the magnitude of S/N ratios significantly, and then they should be adjusted to meet the targets specified by new product planning to meet customers' requirements.

The S/N ratios for measuring the robustness for all kinds of technology development are usually different from each other. It is essential for all enterprises to educate their engineers or technicians in using S/N ratios to measure the functional robustness of their own products and processes. In fact, technology departments should train one quality engineering expert for every 20 engineers or technicians, and this expert should have enough knowledge of robustness to help other engineers and technicians solve any problems they encounter in quality engineering.

It is very important to choose appropriate S/N ratios to solve various quality engineering problems. In addition to the choice of S/N ratios and the knowledge of robustness, measurement technology is another important factor that can affect the efficiency of technology development. From the above, we can see that understanding and applying the concepts of functional robustness are very important to any manufacturing enterprise.

Exercises

[5.1]   A manufacturer decides to buy springs from company A_1 or A_2. The decision will depend on the functional robustness of the springs from these two suppliers. Assume that the extensions of springs are all zero when the load is zero, and choose three levels of loads for the spring force M as follows:

$$M_1 = 0$$
$$M_2 = 10 \text{ gram-force}$$
$$M_3 = 20 \text{ gram-force}$$

The manufacturer will measure the extensions of the springs under these three levels of loads as the objective characteristics. In addition to spring extensions, there are four noise conditions:

N_1 = Low-temperature condition ($-30°C$)
N_2 = High-temperature condition ($+70°C$)
N_3 = Low-temperature condition ($-30°C$) after deterioration test
N_4 = High-temperature condition ($+70°C$) after deterioration test

The experimental data, y_{ij} (extensions of spring measured in millimeters) for the three input signal factors M_i ($i = 1, 2, 3$) and the four compounding noise factors N_j ($j = 1, 2, 3, 4$) are recorded in Table 5.4.

a. Calculate the S/N ratio for the springs made by companies A_1 and A_2 through the weighted linear proportional terms as in Equation (5.4) and compare the functional robustness of A_1 and A_2.

b. Calculate the S/N ratios for A_1 and A_2 by considering only the experimental conditions of N_1 and N_4 and compare the functional robustness of these two types of springs. (In fact, this method might be more efficient than [a].)

c. When the extensions of these springs deviate from the linear proportional part of β by an amount of 5 gram-force under the users' operating conditions, problems will occur and the cost will be $400. Calculate and compare the loss due to functional robustness of the springs from A_1 and from A_2. However, in this case N_1, N_2, N_3, and N_4 should be treated as actual operating noise conditions.

TABLE 5.4 Extensions of springs (y_{ij})

	Company A$_1$				Company A$_2$		
	M_1	M_2	M_3		M_1	M_2	M_3
N_1	-0.3	5.2	10.7	N_1	-0.1	9.8	19.7
N_2	0.3	6.1	11.9	N_2	0.1	10.3	20.4
N_3	1.4	8.3	16.1	N_3	0.4	10.9	21.3
N_4	1.9	11.0	21.3	N_4	0.7	11.5	22.5
Total	3.3	30.6	60.0	Total	1.1	42.5	83.9
	$S_T = 1228.89$				$S_T = 2218.05$		

Hint: Assume that the original S/N ratio is η. The quality loss of a spring can be calculated by the following equation:

$$L = \left(400\right)\left(\frac{1}{5^2}\right)\left(\frac{1}{\eta}\right)$$

[5.2] In the technological development of an infrared soldering process, both factors A and B have three levels; factors C, D, E, and F all have two levels. All six controllable factors are assigned to an L_9 array. However, C and D are combined into three levels, C_1D_1, C_1D_2, and C_2D_1. Similarly, E and F are combined into three levels, E_1F_1, E_1F_2, and E_2F_1. In addition to these six controllable factors, there are three compound noise conditions: N_1 = most negative, N_2 = standard, and N_3 = most positive. Let the threshold values for the nonsoldering conditions under the three compound noise conditions be x_1, x_2, and x_3. Also, let the threshold values for bridging problems (because of overheating) under these three compound noise conditions be y_1, y_2, and y_3. The S/N ratio

TABLE 5.5 L$_9$ orthogonal array and operating window

Test No.	A 1	B 2	(CD) 3	(EF) 4	η_x	η_y	η
1	1	1	1	1	−2.0	−1.4	−3.4
2	1	2	2	2	1.4	−6.1	−4.7
3	1	3	3	3	3.2	−2.6	0.6
4	2	1	2	3	1.0	3.3	4.3
5	2	2	3	1	−0.5	3.7	3.2
6	2	3	1	2	−10.3	3.7	−6.6
7	3	1	3	2	2.7	3.3	6.0
8	3	2	1	3	0.0	−1.8	−1.8
9	3	3	2	1	1.5	−2.2	−0.7

for the operating window of the soldering process is calculated by the following equations:

$$\eta = \eta_x + \eta_y$$

$$\eta_x = -10 \log\left(\frac{x_1^2 + x_2^2 + x_3^2}{3}\right)$$

$$\eta_y = -10 \log\left[\frac{1}{3}\left(\frac{1}{y_1^2} + \frac{1}{y_2^2} + \frac{1}{y_3^2}\right)\right]$$

The data in Table 5.5 show the amount of electrical power (watts) needed for the soldering process, which is the objective experimental characteristic:

a. Find the optimal levels for these six controllable factors. What is the gain in functional robustness under the optimal conditions compared with the current operating conditions, A_2, B_2, C_1, D_1, E_1, and F_1? In this question, all main effects must be considered.

b. Under the optimal condition, estimate the values of η_x and η_y. Also, estimate the equivalent lower threshold values $\bar{x}$ and the equivalent higher threshold values $\bar{y}$. Also, calculate the power (in watts) needed for the optimal condition. All six controllable factors should be taken into consideration in this question. The equivalent threshold values for $\bar{x}$ and $\bar{y}$ can be calculated by the following equations:

$$\text{Estimated value of } \eta_x = -10 \log(\bar{x})^2$$

$$\text{Estimated value of } \eta_y = -10 \log\left(\frac{1}{y}\right)^2$$

SOLUTIONS TO EXERCISES _____

1.1.   The ideal function of a spring is to behave such that $F = kx$, where F is the force, k is the stiffness, and x is the displacement.

   The objective function of a real spring will have non-linearity due to material properties, manufacturing, temperature, age and the like (call these properties p_i)

   The real function becomes

$$F = kx + f(p_i).$$

The spring should be designed so that

$$|f(p_i)| \leq \Delta$$

where Δ is the tolerance, for all properties within the operating range, for a given Δ.

2.1a.   For 3 materials, loss is calculated as:

$$L = \text{Price} + \frac{A_0}{\Delta_0^2} \left(b^2 \sigma_{\text{temp}}^2 + \frac{T^2}{3} \beta^2 \right)$$

$$A_1: L = 1.8 + \frac{280}{10^2} \left[\left(0.08^2 \right) \left(15^2 \right) + \left(\frac{10^2}{3} \right) \left(0.15^2 \right) \right]$$

$$= \$7.93$$

$$A_2: L = 3.5 + \frac{280}{10^2}\left[\left(0.03^2\right)\left(15^2\right) + \left(\frac{10^2}{3}\right)\left(0.06^2\right)\right]$$

$$= \$4.40$$

$$A_3: L = 6.3 + \frac{280}{10^2}\left[\left(0.01^2\right)\left(15^2\right) + \left(\frac{10^2}{3}\right)\left(0.05^2\right)\right]$$

$$= \$6.60$$

Material	Price ($)	Quality Loss	Total Loss ($)
A_1	1.80	613	7.93
A_2	350	91	4.41
A_3	630	30	6.60

A_2 is the best material.

2.1b. The optimal tolerance Δ is found from

$$\frac{A}{\Delta^2} = \frac{A_0}{\Delta_0^2}$$

where A is the manufacturer's loss if the product is defective. (For material A_2, $A = \$3.50$.)

$$\Delta = \sqrt{\frac{A}{A_0}}\,\Delta_0 = \left(\sqrt{\frac{3.5}{280}}\right)\left(10\right) = 1.12\%$$

3.1. If after shipment the car does not meet the spec $\Delta_0 = 0.48$ gal/km the cost is $A_0 = 50000$ ¥. Inside the plant, the adjusting cost is $A = 6000$ ¥.

$$\phi = \sqrt{\frac{A_0}{A}} = \sqrt{\frac{50000\ ¥}{6000\ ¥}} = 2.89$$

The manufacturer's spec should be

$$\Delta = \frac{\Delta_0}{\phi} = \frac{0.48}{2.89} = 0.17 \text{ g/km}$$

3.2.   The functional limits of the output voltage are $\Delta_0 = 50\%$. The cost for exceeding the limits is $A_0 = \$180$ while the cost of adjustment is $A = \$0.50$

$$\phi = \sqrt{\frac{A_0}{A}} = \sqrt{\frac{180}{0.5}} = 18.97$$

$$\Delta = \frac{\Delta_0}{\phi} = \frac{50}{18.97} = 2.64\%$$

3.3.   The tolerance is determined from:

$$\Delta = \sqrt{\frac{A}{A_0}} \left(\frac{\Delta_0}{\beta} \right)$$

Here $A_0 = 0.20$ (scrapping cost of network)

$A = 0.50$ (scrapping of resistor)

$\Delta_0 = 5$ mV (tolerance on network)

$\beta = 0.8$ mV (network deviation for

resistor deviation of 1 mV)

$$\Delta = \sqrt{\frac{0.50}{0.20}} \left(\frac{5 \text{ mV}}{0.8 \text{ mV}} \right) = 3.125 \text{ mV}$$

3.4a.   The tolerance wear rate specifications are found from:

$$\Delta^* = \sqrt{\frac{3A^*}{A_0}} \frac{\Delta_0}{\beta T}$$

Here $A^* = \$0.50$ (cost of defective product)

$A_0 = \$200$ (scrapping of product)

$\Delta_0 = 120$ μm (tolerance limit on product)

$\beta = 1$ μm (component deviation causes some product deviation)

$T = 10$ years

$$\Delta^* = \sqrt{\frac{(3)(0.50)}{200}} \frac{120}{1.10} = 1.04 \text{ μm/year}$$

3.4b. Follow-up: Preventive maintenance

none: $L = \left(\frac{200}{120^2}\right)\left(\frac{10^2}{3}\right)\left(15^2\right) = \$104.16$

once: $L = 30 + \left(\frac{200}{120^2}\right)\left(\frac{5^2}{3}\right)\left(15^2\right) = \$56.04$

Preventive maintenance should be conducted once every five years.

4.1a. Using equations 4.4 and 4.5, savings from the optimal control system are found as:

$$(L_0 - L)(\# \text{ units/hour})(\# \text{ hours})$$

with

$\Delta = 40$ μm

$A = \$1.20$

$n_0 = 120$ units

$D_0 = 10$ μm

B = $1.20
l = 2
C = $3
u_0 = (20 hours)(120 units/hour) = 2400 units

Current $L_0 = \dfrac{B}{n_0} + \dfrac{C}{u_0} + \dfrac{A}{\Delta^2}$

$$* \left[\frac{D_0^2}{3} + \left(\frac{n_0 + 1}{2} + l\right)\frac{D_0^2}{u_0}\right]$$

$$= \frac{1.20}{120} + \frac{3}{2400} + \frac{1.20}{40^2}$$

$$* \left[\frac{10^2}{3} + \left(\frac{120 + 1}{2} + 2\right)\frac{10^2}{2400}\right]$$

$$= 0.0100 + 0.0013 + 0.0250 + 0.0020$$

$$= \$0.0383, \text{ round to } \$0.038$$

Optimum $n = \sqrt{\dfrac{2u_0B}{A}}\left(\dfrac{\Delta}{D_0}\right) = \sqrt{\dfrac{(2)(2400)(1.20)}{1.20}}\left(\dfrac{40}{10}\right)$

$$= 277 \text{ units, round to 240 units}$$

$$D = \left[\left(\frac{3C}{A}\right)\left(\frac{D_0^2}{u_0}\right)\left(\Delta^2\right)\right]^{1/4}$$

$$= \left[\left(\frac{3 \cdot 3.00}{1.20}\right)\left(\frac{10^2}{2400}\right)\left(40^2\right)\right]^{1/4}$$

$$= 4.73 \ \mu m, \text{ round to } 5.0 \ um$$

$$L = \left(\frac{B}{n} + \frac{C}{u} + \frac{A}{\Delta^2}\right)\left[\frac{D^2}{3} + \left(\frac{n + 1}{2} + l\right)\frac{D^2}{\mu}\right]$$

$$= \frac{1.20}{277} + \frac{3.00}{537} + \frac{1.2}{40^2}$$

$$* \left[\frac{4.73^2}{3} + \left(\frac{277 + 1}{2} + 2 \right) \frac{4.73^2}{537} \right]$$

$$= 0.0043 + 0.0056 + 0.0056 + 0.0044$$

$$= \$0.0199, \text{ round to } \$0.02$$

$$u = u_0 \left(\frac{D^2}{D_0^2} \right) = 2400 \left(\frac{4.73^2}{10^2} \right)$$

$$= 537 \text{ units, round to } 600 \text{ units}$$

Savings: $(0.038 - 0.02)(120)(2000) = \$4320/\text{year}$.

4.1b. Process capability index $C_p = \dfrac{2\Delta}{6\sigma}$

$$\sigma_{\text{current}} = \sqrt{\frac{D_0^2}{3} + \left(\frac{n_0 + 1}{2} + l \right) \left(\frac{D_0^2}{u_0} \right)}$$

$$= \sqrt{\frac{10^2}{3} + \left(\frac{120 + 1}{2} + 2 \right) \left(\frac{10^2}{2400} \right)} = 5.99$$

$$C_{p\text{current}} = \frac{(2)(40)}{(6)(5.99)} = 2.22$$

$$\sigma \text{ optimal} = \sqrt{\frac{D^2}{3} + \left(\frac{n + 1}{2} + l \right) \left(\frac{D^2}{u} \right)}$$

$$= \sqrt{\frac{5^2}{3} + \left(\frac{240 + 1}{2} + 2 \right) \left(\frac{5^2}{600} \right)} = 3.67$$

$$C_{p\text{optimal}} = \frac{(2)(40)}{(6)(3.67)} = 3.63$$

4.1c.   Man-hours = (total operation hours) −
[(2/60)(# measurements/year) +
(8/60)(# adjustments/year)]

Original man-hours = 2000 − [(2/60)(2000/2) +
(8/60)(2000/20)] = 1920 hours

Optimal man-hours = $2000 - \left[\left(2/60\right)\left(2000/2\right) + \left(8/60\right)\left(\dfrac{2000}{600/120}\right)\right]$ = 1913.33 hours

Here $\Delta = 3.5$ μm
$A = \$3$
$B = \$15$
$D_0 = 0.5$ μm
$l = 0$
$C = \$30$
$u_0 = 5$ batches
$n_0 = 1$ batch

Production rate: 6 batches/day for 250 days/year.

4.2a.   Production unit is one batch.

$A = (\$3/\text{unit})(800 \text{ units/batch})$
$\quad = \$2400/\text{batch}$

Current $L_0 = \dfrac{B}{n_0} + \dfrac{C}{u_0} + \dfrac{A}{\Delta^2}$

$$* \left[\dfrac{D_0^2}{3} + \left(\dfrac{n_0 + 1}{2} + l\right)\dfrac{D_0^2}{u_0}\right]$$

$$= \frac{15}{1} + \frac{30}{5} + \frac{2400}{3.5^2}$$

$$* \left[\frac{0.5^2}{3} + \left(\frac{1+1}{2} + 0 \right) \frac{0.5^2}{5} \right]$$

$$= 15 + 6 + 16.3 + 9.80 = \$47.13$$

$$\text{Optimum } n = \left(\sqrt{\frac{2\, u_0 B}{A}} \right) \left(\frac{\Delta}{D_0} \right)$$

$$= \left(\sqrt{\frac{(2)(5)(15)}{2400}} \right) \left(\frac{3.5}{.5} \right)$$

$$= 1.7 \text{ batches, round to } 2$$

$$D = \left[\frac{3C}{A} \frac{D_0^2}{u_0} \Delta^2 \right]^{1/4}$$

$$= \left[\left(\frac{3 \times 30}{2400} \right) \frac{0.5^2}{5} 3.5^2 \right]^{1/4}$$

$$= 0.39 \ \mu\text{m, round to } 0.4 \ \mu\text{m}$$

$$u_0 = u_0 \left(\frac{D^2}{D_0^2} \right) = 5 \left(\frac{0.4^2}{0.5^2} \right) = 3.2$$

$$L = \frac{B}{n} + \frac{C}{u} + \frac{A}{\Delta^2}$$

$$* \left[\frac{D^2}{3} + \left(\frac{n+1}{2} + l \right) \frac{D^2}{u} \right]$$

$$= \frac{15}{2} + \frac{30}{3.2} + \frac{2400}{3.5^2}$$

$$* \left[\frac{0.4^2}{3} + \left(\frac{2+1}{2} + 0 \right) \frac{0.4^2}{3.2} \right]$$

$$= 7.5 + 9.38 + 195.92\ (0.05 + 0.075)$$

$$= \$42.02$$

$$\text{Gain} = (L_0 - L)(\# \text{ batches/day})(\# \text{ day/year})$$

$$= (47.13 - 42.02)(6)(250)$$

$$= \$7665$$

4.2b.
$$L_0 = 47.13 + \left(\frac{2400}{3.5^2}\right)\left(0.3^2\right) = \$64.76$$

Optimum $n = 2$, $D = 0.4$

$$L = \frac{22.50}{2} + \frac{30}{3.2} + \frac{2400}{3.5^2}$$

$$* \left[\frac{0.4^2}{3} + \left(\frac{3}{2} + 0\right)\frac{0.4^2}{3.2} + 0.15^2\right]$$

$$= \$50.18$$

$$\text{Gain} = (L_0 - L)(\# \text{ batch/day})(\# \text{ day/yr})$$

$$= [(64.76 - 50.18)/\text{batch}](6)(250)$$

$$= \$21,870/\text{year}$$

Conclusion: Gain indicates new method is better.

5.1a. Following Equations 5.4–5.7, and extending them from 2 to 4 noise conditions, we can calculate the S/N ratio (η):

Company A_1

L is calculated using values of M and y_{ij} given in Table 5.4.

$$L_1 = M_1y_{11} + M_2y_{12} + M_3y_{13}$$

$$= 0(-0.3) + 10(5.2) + 20(10.7)$$

$$= 266$$

$$L_2 = M_1y_{21} + M_2y_{22} + M_3y_{23}$$

$$= 0(0.3) + 10(6.1) + 20(11.9)$$

$$= 299$$

$$L_3 = M_1y_{31} + M_2y_{32} + M_3y_{33}$$

$$= 0(1.4) + 10(8.3) + 20(16.1)$$

$$= 405$$

$$L_4 = M_1y_{41} + M_2y_{42} + M_3y_{43}$$

$$= 0(1.9) + 10(11.0) + 20(21.3)$$

$$= 536$$

$$r = 4(M_1^2 + M_2^2 + M_3^2) = 4(0^2 + 10^2 + 20^2) = 2000$$

$$S_{N*\beta} = \frac{(L_1 - L_2)^2 + (L_1 - L_3)^2 + (L_1 - L_4)^2 + (L_2 - L_3)^2 + (L_2 - L_4)^2 + (L_3 - L_4)^2}{r}$$

$$= 88.94$$

$$S_\beta = \frac{(L_1 + L_2 + L_3 + L_4)^2}{r}$$

$$= \frac{(266 + 299 + 405 + 536)^2}{2000}$$

$$= 1134.02$$

From Table 5.4

$$S_T = y_{11}^2 + y_{12}^2 + y_{13}^2 + y_{21}^2 + y_{22}^2 + y_{23}^2$$

$$= 1228.89$$

$$S_e = S_T - S_\beta - S_N = 5.93$$

$$V_e = \frac{S_e}{4K - 4} = \frac{5.93}{(4)(3) - 4} = 0.74$$

$$V_N = \frac{S_{N*\beta} + S_e}{4K - 1} = \frac{88.94 + 5.93}{(4)(3) - 1} = 8.62$$

Company A_2

$$L_1 = 0(-0.1) + 10(9.8) + 20(19.7) = 492$$

$$L_2 = 0(0.1) + 10(10.3) + 20(20.4) = 511$$

$$L_3 = 0(0.4) + 10(10.9) + 20(21.3) = 535$$

$$L_4 = 0(0.7) + 10(11.5) + 20(22.5) = 565$$

$$r = 2000$$

$$S_\beta = \frac{(492 + 511 + 535 + 565)^2}{2000} = 2211.3$$

$$S_{N*\beta} = 5.97$$

$$S_T = 2218.05$$

$$S_e = 0.78$$

$$V_e = 0.098$$

$$V_N = 0.614$$

Analysis of Variance (ANOVA)

Source	DOF	A_1 S	A_1 V	A_2 S	A_2 V
β	1	1134.02		2211.30	
$N*\beta$	3	88.94		5.97	
e	8	5.93	0.74	0.78	0.098
$N*\beta + e$	11	94.87	8.62	6.75	0.614

$$\eta = 10 \log \frac{1}{r} \left(\frac{S_\beta - V_e}{V_N} \right)$$

$$\eta(A_1) = 10 \log \frac{\dfrac{1}{2000}(1134.02 - 0.74)}{8.62}$$

$$= -11.8 \text{ decibels}$$

$$\eta(A_2) = 10 \log \frac{\dfrac{1}{2000}(2211.3 - 0.098)}{0.614}$$

$$= 2.6 \text{ decibels}$$

Gain $= 14.4$ decibels

The robustness of A_2 is 27.38 times ($10^{\frac{14.37}{10}}$) better than the robustness of A_1.

5.1b. Consider only N_1 and N_4

$$r = 2(M_1^2 + M_2^2 + M_3^2)$$

$$= 1000$$

Company A_1

$$S_\beta = \frac{(L_1 + L_4)^2}{r}$$

$$= \frac{(266 + 536)^2}{1000}$$

$$= 643.2$$

$$S_{N*\beta} = \frac{(L_1 - L_4)^2}{r}$$

$$= \frac{(266 - 536)^2}{1000}$$

$$= 72.9$$

$$S_T = y_{11}^2 + y_{12}^2 + y_{13}^2 + y_{41}^2 + y_{42}^2 + y_{43}^2$$

$$= (-0.3)^2 + 5.2^2 + 10.7^2 + 1.9^2 + 11^2 + 21.3^2$$

$$= 719.92$$

$$S_e = S_T - S_\beta - S_{N*\beta}$$

$$= 719.92 - 643.2 - 72.9$$

$$= 3.82$$

$$V_e = \frac{S_e}{2K - 2} = \frac{3.82}{(2)(3) - 2} = 0.96$$

$$V_N = \frac{S_N + S_e}{2K - 1} = \frac{72.9 + 3.82}{(2)(3) - 1} = 15.34$$

Company A_2

$$S_\beta = 1117.25$$

$$S_{N*\beta} = 5.33$$

$$S_T = 1123.13$$

$$S_e = 0.55$$

$$V_e = 0.14$$

$$V_N = 1.176$$

$$\eta(A_1) = 10 \log \frac{\dfrac{1}{1000}(643.20 - 0.96)}{15.34}$$

$$= -13.8 \text{ decibels}$$

$$\eta(A_2) = 10 \log \frac{\dfrac{1}{1000}(1117.25 - 0.14)}{1.176}$$

$$= -0.2 \text{ decibels}$$

$$\text{Gain} = 14 \text{ decibels}$$

5.1c. $A_1: \dfrac{1}{r}\left(\dfrac{S_\beta - V_e}{V_N}\right) = 0.0657$

$A_2: \dfrac{1}{r}\left(\dfrac{S_\beta - V_e}{V_N}\right) = 1.80$

$A_1: L = \left(\dfrac{400}{5^2}\right)\left(\dfrac{1}{0.0657}\right) = \$243.53$

$$A_2: L = \left(\frac{400}{5^2}\right)\left(\frac{1}{1.80}\right) = \$8.87$$

Gain = \$234.66

5.2   Calculate η_x, η_y, and η using Table 5.5. For example:

$$(A_1)\eta_x = -2.0 + 1.4 + 3.2 = 2.6$$

$$\eta_y = -1.4 - 6.1 - 2.6 = -10.1$$

$$\eta = \eta_x + \eta_y = -7.5$$

and so on.

Calculate the total average by adding all factors and dividing by 4.

Factorial Effect Table (Total)

	η_x	η_y	η
A_1	2.6	-10.1	-7.5
A_2	-9.8	10.7	0.9
A_3	4.2	-0.7	3.5
B_1	1.7	5.2	6.9
B_2	0.9	-4.2	-3.3
B_3	-5.6	-1.1	-6.7
C_1D_1	-12.3	0.5	-11.8
C_1D_2	3.9	-5.0	-1.1
C_2D_1	5.4	4.4	9.8
E_1F_1	-1.0	0.1	-0.9
E_1F_2	-6.2	0.9	-5.3
E_2F_1	4.2	-1.1	3.1
Total Average	-3.0	-0.1	-3.1

5.2a. The optimum combination of factors are those with the largest values of η:

A_3, B_1, C_2D_1, E_2F_1

Estimate:

$$\text{Optimum:} \quad \hat{\mu} = \frac{3.5 + 6.9 + 9.8 + 3.1}{3}$$

$$\left(-3\right)\left(\frac{-3.1}{9}\right) = 8.8$$

$$\text{Current:} \quad \hat{\mu} = \frac{0.90 - 3.3 - 11.8 - 0.9}{3}$$

$$\left(-3\right)\left(\frac{-3.1}{9}\right) = -4.0$$

$$\text{Gain:} \quad 8.8 - (-4.0) = 12.8 \text{ (decibel)}$$

5.2b. $\bar{x} = \sqrt{10^{-0.617}} = 0.49$ (W)

$\bar{y} = \sqrt{10^{-0.263}} = 0.74$ (W)

Optimum level of power $= \dfrac{1}{2}(0.49 + 0.74) = 0.61$

Index